CLASSIC AND SPORTSCAR

TRIUMPH SPITFIRE AND GT6 FILE

CLASSIC AND SPORTSCAR

TRIUMPH SPITFIRE AND GT6 FILE

GRAHAM
ROBSON

Series Editor Mark Hughes

TEMPLE PRESS

First published 1988 by
Temple Press
an imprint of
The Hamlyn Publishing Group Ltd
a division of
The Octopus Group PLC
Michelin House, 81 Fulham Road
London SW3 6RB
and distributed for them by
Octopus Distribution Services Ltd
Rushden, Northants NN10 9RZ

Produced by Bay View Books Ltd
Bideford, Devon EX39 2QE
Designed by Gerrard Lindley
Colour photography by Paul Debois
Picture research by Jon Pressnell
Typeset by Lens Typesetting, Bideford
Reproduction by Isca Reproductions, Exeter

ISBN 0 600 55780 4
Printed by Mandarin Offset in Hong Kong

Contents

A Backbone to Success

Although Triumph was a famous Coventry name, which built its first car in 1923, it suffered several financial crises in the 1930s, and was finally bought up by the Standard organization in 1944. The pre-war Triumph concern had built several sporty cars, but never challenged MG in the two-seater sports car market; up to 1945 Standard had *never* produced a sports car.

The 1950s and 1960s were the years in which the Standard-Triumph organization really surged ahead, building not only more and more touring cars, but an attractive range of sports cars, beginning with the TR2. When you consider that there was no established sports-car pedigree – either Standard *or* Triumph – this was a remarkable achievement. Like MG, Triumph impressed the 1950s enthusiasts by first producing a medium-sized sports car, and then followed it up with a smaller machine. In the case of MG, it was MGA first and Midget later; with Triumph it was TR2 and (10 years later) the Spitfire.

The Spitfire's direct ancestor was the Triumph Herald family, which was introduced in 1959. Until the Spitfire arrived the most sporting version was the coupé, which had 2+2 seating. It was heavier, larger and less powerful than the Spitfire would be.

When the Herald range was introduced in 1959, the coupé had this twin-carb 948cc engine as standard. Clearly this was the immediate forerunner of the 1147cc engine used three years later in the Spitfire.

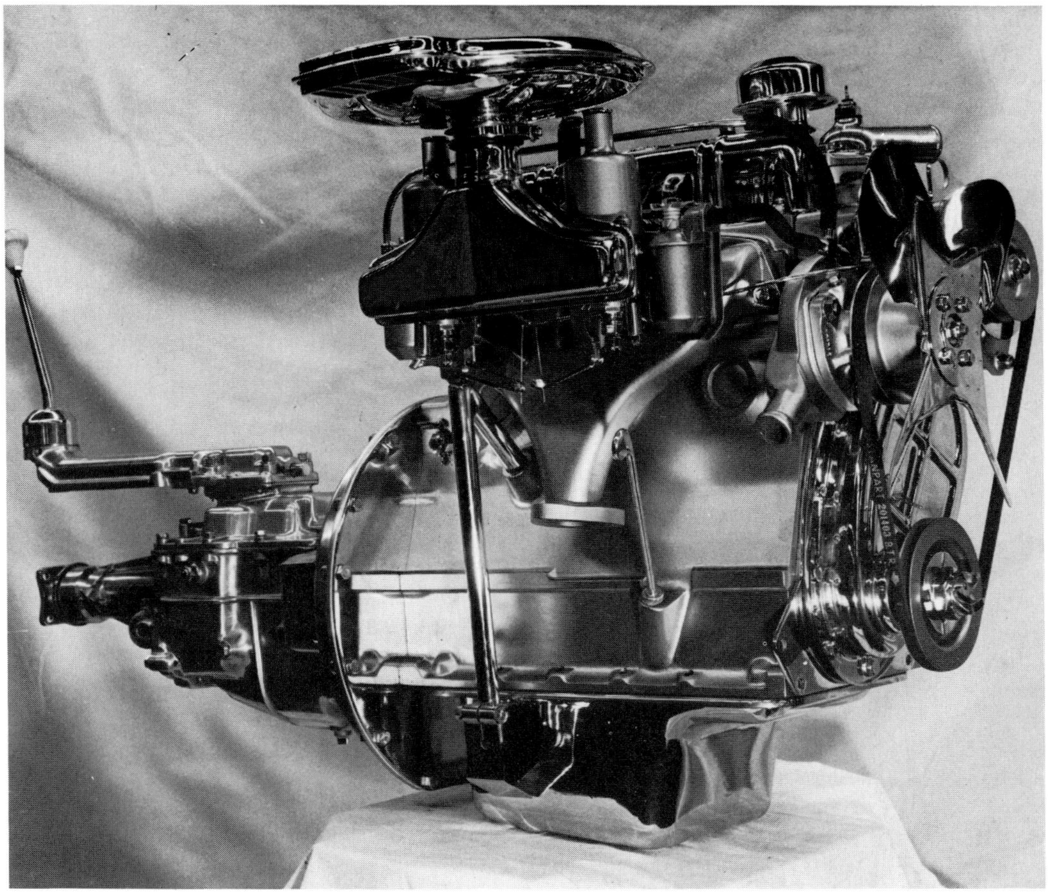

Because of the unique design of its front suspension, the Herald had a phenomenal turning circle, as this publicity shot emphasizes. The Spitfire, with its shorter wheelbase, was even more nimble.

The Spitfire

Everyone – and that included Triumph's deadliest rivals at Abingdon – had to agree that the Spitfire was a very pretty little car. But like most such quantity production sports cars of its day, under the sleek styling there were surprisingly mundane chassis and mechanical components. The truth of the matter was that the 90mph Spitfire of 1962 had begun to take shape several years earlier. Its general layout dated back to 1957 and the concept of the Triumph Herald, while the engine and transmission components of the Herald/Spitfire range all had their origins in the dumpy little Standard Eight saloon of 1953. Purely in passing, it's worth recalling that the Standard Eight, as conceived, *might* just have had a side-valve engine and a three-speed gearbox if its original sponsors had had their way – and if this had been approved the Eight would surely have been a failure, and the Spitfire would never even have been invented!

The Spitfire arrived at a time when Standard-Triumph was not only in corporate turmoil (the takeover by Leyland had been formalized in 1961), but when some of its products were, to say the least, controversial.

The Spitfire however, always had a successful career. The Americans loved it and, according to technical chief Harry Webster, they very rarely complained about the rear suspension. It sold well in many countries (and particularly in the United States), mainly because it was prettier and much more roomy than the Austin-Healey Sprite/MG Midget rivals. Even though the Herald/Vitesse cars from which it had been developed disappeared at the beginning of the 1970s, the Spitfire stayed in production until the summer of 1980, by which time no fewer than 314,342 cars had been built.

The story really dates from 1951, when technical director Ted Grinham and his chassis designer Harry Webster started to lay out a new small car (logically enough this was coded 'SC' – for Small Car). This was the first true small post-war Standard, and would have to be a sound 'building block' for the future. That car eventually became the Standard Eight/Ten range, first seen in 1953/54, and was *all* new – engine, transmission, axle, suspension and bodyshell. The engine, at that time, had 803cc and 28bhp, or 948cc and 33bhp, depending on the model. More by chance than by original intention, the engine was found to have a great deal of built-in 'stretch', for it was eventually enlarged to 1493cc. The gearbox too, was stronger than expected, and eventually put up with the 75bhp of the Spitfire Mk 3 without protest.

Then, in 1959, came the Michelotti-styled Triumph Herald, whose separate chassis and multi-section bodyshell (and a multitude of styles) were inspired by a desire to assemble it abroad from simple jigging, and because Standard's previous suppliers, Fisher & Ludlow, had been taken over by BMC, which was no longer willing to support the activities of a rival. There were always doubts about Herald quality and pricing (it was too expensive), and controversy about the use of swing-axle independent rear suspension.

A sports car derivative of the Herald was soon under discussion, as a tripartite 'invention' of Harry Webster (technical chief since 1957), George Turnbull (in line to become Standard-Triumph's production chief in the 1960s), and the resourceful Italian stylist Giovanni Michelotti, who had been under contract to Standard-Triumph since 1957; managing director Alick Dick was only advised when the idea had taken root and the first schemes had been laid. As with other Triumph projects of the day, it gained a project name – in this case 'Bomb'. Webster soon settled on a wheelbase of 83in, (which was 8½in less that of the Herald saloon) and shipped a 'cut-and-shut' Herald frame out to Turin to let Michelotti get on with the styling job.

Michelotti's style, which was influenced by Webster on his regular visits to Turin, was complete in mid-1960, and the prototype

top right

Michelotti built this, the original Spitfire prototype, in 1960, using the running gear of a Triumph Herald coupé. Except in detail, the external lines were not changed before mass-production tooling was laid down.

bottom right

Michelotti's cheap and cheerful Spitfire prototype facia was fortunately not adopted for the production cars. In fact all *the dials and switches had been lifted from the Herald coupé which he cannibalized to produce the original car.*

To get the show on the road in 1960/1961, much development running was done on a short-wheelbase Herald with only two seats. Note the short doors, betrayed by the discontinuity in the styling crease along the flanks.

bodyshell, by this time on a proper chassis, was sent back to Standard-Triumph in Coventry in October. And there, after brief evaluation tests, it stayed under wraps for the next year . . .

By that time Standard-Triumph's finances had collapsed into disarray, but the company would not formally be bailed out and rescued by Leyland Motors until 1961. Because the company was losing so much money, no-one dared propose introducing a small sports car at first, and it was not until Standard-Triumph's new chief from Leyland, Stanley Markland, saw the prototype (which had been under a dust sheet for a long time) that production approval was given in July 1961.

Tooling was quickly commissioned in time for an October 1962 launch, by which time the first production cars were being completed at the Canley factory – without today's complex legislative requirements, things moved much faster in those days! Bodyshells were produced by the Forward Radiator Co. of Birmingham (this was a Standard-Triumph subsidiary), chassis frames by a 'heavy metal basher' in the West Midlands, while Standard-Triumph machined and assembled its engines and transmissions in factories spread all around Coventry.

The 'Spitfire' name came along while prototypes were in development, but there does not seem to have been *any* connection with the famous fighter aircraft, though publicity pictures were eventually taken with car posed in front of 'plane. It is not recorded if the makers of the 'plane ever objected to its name being 'stolen' for use on a car.

In general the Spitfire was a very well-conceived car, and most people who drove it loved it. I owned a Spitfire Mk 1 for a couple of years, and loved the spacious cockpit, the styling, and the eager performance. My car had lowish rear suspension and handled tolerably well, but there was no doubt that many Spitfires had very precarious rear-end grip, particularly on worn tyres and in slippery conditions.

There were always many good points to enjoy though, such as the amazing agility in traffic (that turning circle was remarkably compact – so much so that front wheel tyre scrub could be a serious problem), the crisp gearchange of a car not yet worn out, the willing roar of the engine, and the looks. Many people never took to the alternative, the narrow, rather straight-line looks of the contemporary Spridget, or its sliding side-curtains. The rivals always had very similar performance, so it was the better specification which made the Spitfire so attractive.

The Spitfire, as announced in 1962, was a compact and very pretty little sports car. This Autocar *cutaway drawing shows how the modified Herald running gear had been installed.*

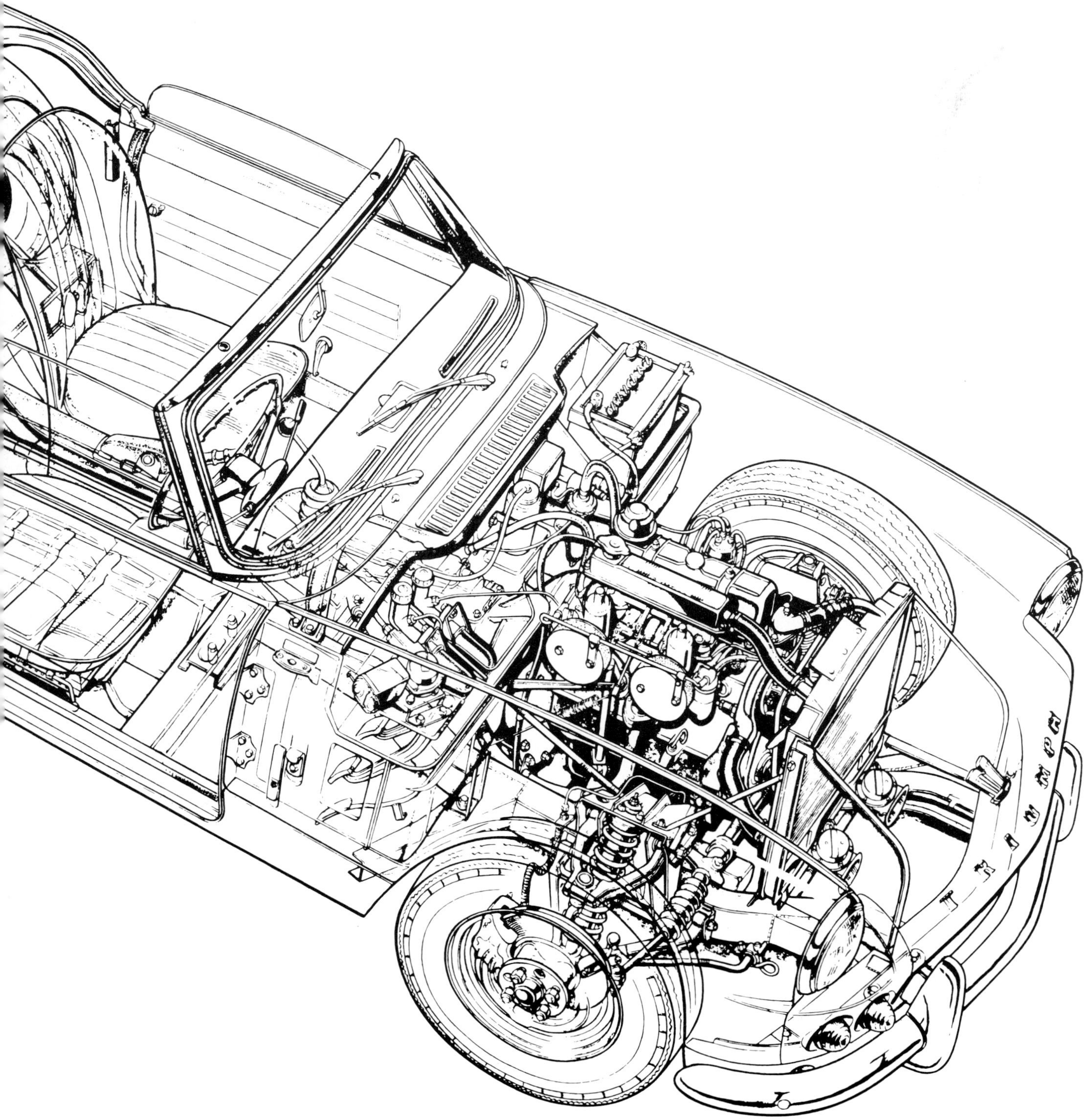

QJ J
SPITFIRE

PAE 565F

The final type of UK market Spitfire, the 1500 model, launched at the end of 1974. The body had been completely reskinned a few years earlier for the Mk IV,

top left

Spitfire car posed in front of Spitfire fighter. This was an early production car, captured in 1962.

bottom left

By 1968 the Spitfire had become Mk 3, complete with raised front bumper, a 75bhp 1.3-litre engine, and a plusher interior. The wire-spoke wheels were optional equipment.

Overdrive was soon offered, and this was a desirable fitting, for no other small sports car had ever offered the feature. The result was relaxed motorway cruising (motorways were rare, and quite new, in the UK at the time) with astonishing economy. It was also nice to know that the cars could be turned into formidable little race and rally cars if you threw enough money at them – as the factory proved at Le Mans and in the Tour de France and Alpine rallies – for it meant that the engine had hidden reserves, even in its 'stretched' state.

Many improvements made in later years were inspired by work also done on the Herald saloons, and on the six-cylinder derivatives, the Vitesse family car and the GT6 coupé. If not actually overdue, they were certainly much appreciated when they eventually arrived. All-synchromesh gears for the Spitfire arrived in 1970 (the GT6 always had such a transmission), while the tamed rear suspension was phased in at the same time. The more torquey engine – the 1493cc unit of 1974 – was so good that it was also fitted to MG Midgets manufactured from that time.

Some features were present right from the start, and were always welcome, such as the windproof and waterproof interior, especially in optional hardtop form. (The Herald's water leaking problems never cropped up on Spitfires because of the sports car's different type of body construction). There was the splendid all-round visibility from the cockpit (the windscreen, after all, was shared with that used in the TR4/TR4A/TR5/TR250/TR6 family of sports cars), and the ideal engine bay access provided by the lift-up front end, which combined bonnet with wings *and* inner wheel arches.

The steering, although low geared, was always finger-light and accurate, the brakes with front discs were good (and even fade-free if you didn't expect too much of them), and the potential fuel economy could be a lot better than one would expect. Even the insurance companies eventually learned to be quite relaxed about the Spitfire.

Although the Spitfire remained in production for 18 years, and sold well right up to the end, it never received all the development attention that it deserved. Some types of Spitfire were definitely better than others. Not surprisingly the car suffered most from neglect in the early years of British Leyland, when Standard-Triumph was not only struggling to keep its head above the political waters, but its engineers were also having to deal with new cars as diverse as the Stag and the Toledo, the first of the TR7s, and the first of the Dolomites.

There isn't much doubt that the best Spitfires to buy in the late 1980s are those which handle well and go fastest. Almost automatically this means that the cars of the 1970s, and preferably the 1500s in UK trim, are the most desirable. Of the earlier models, the Mk 3s are well-liked because they are quicker than their ancestors *and* the Mk IVs which followed, but you should also be sure that a particular car has relatively secure handling: negative camber rear wheels and radial ply tyres help a lot.

The best Spitfires to buy are also those which come loaded up with factory options. The most obvious, and most frequently found, are the detachable hardtop, the overdrive, and the centre-lock wire spoke wheels; the wheels were not offered after 1973 because demand had dropped away, though some later cars have been converted to that specification in more recent years.

Big sister – the GT6

Although the GT6 was clearly a development of the Spitfire, it had an entirely different type of character. Whereas the Spitfire was always a cheap and cheerful four-cylinder engined open car, the GT6 was a closed coupé/ hatchback with a smooth six-cylinder engine, and an altogether more sophisticated image. As we have already seen, the Herald came first, and the Spitfire followed it. In the meantime, though, Standard-Triumph had evolved a six-cylinder version of the four-cylinder engine, and had been dabbling with a fastback coupé version of the Spitfire. New readers start here . . .

Works Spitfires raced at Le Mans in 1964 and 1965. The fastback roofs were glassfibre mouldings taken off the GT6 pre-prototype.

top right

Two of the works rally cars ready to leave Coventry for the Geneva Rally of 1964, where they finished second and fifth overall. The glassfibre roofs were the same shape as those of the race cars, but with extra spare wheel 'luggage slots' in the tail.

bottom right

The Triumph 2000 saloon was announced in 1963, with a 95bhp version of the company's 1998cc engine. It only needed minor modifications to make it suitable for the GT6, which followed in 1966.

GB
ADU 7B
GB
ADU 6B

LUCAS

In the early 1960s, the Webster-Turnbull-Dick 'think-tank' mated the longer six-cylinder engine to the Herald's chassis, called this project 'Atom', and refined it into the new Vitesse, which was launched in 1962, ahead of the Spitfire. There was definitely no intention of building a six-cylinder version of the Spitfire at that time. In the next few years, however, a new car evolved, influenced by a Michelotti 'Spitfire GT' prototype, by the need to improve the Vitesse, and by the use of the 'Spitfire GT' body shape on the 'works' racing and rallying Spitfires.

First, in 1963, Michelotti produced a one-off 'Spitfire GT' prototype, with a fastback/hatchback coupé roof and a much more luxuriously-trimmed interior. This car, however, was slower and more thirsty than the normal Spitfire, so during 1964 it was re-engined with a 1.6-litre six-cylinder engine. At that time it had only 77bhp, and could reach 90-95mph. Later the same car was used as the 'body mould' for all the 'works' racing and rally Spitfires to be given glass-fibre coupé roof styles, which meant that the public saw the future GT6's shape even before Standard-Triumph had decided to build such a car.

It was only in 1965, when the rest of the British motor industry got to know of BMC's plans, that the GT6 was born. BMC, it was said, had hired Pininfarina to help MG produce a fastback/hatchback version of the MGB model; such a car, indeed, was introduced in the autumn of 1965 and sold immediately very well. Before then, the Triumph prototype didn't really seem to have a market niche – but now a new one, and a new rival, was opening up before its very headlamps.

From that point onwards, a series of happy coincidences occurred. A lot of development was already being done to improve the Vitesse, so work then went ahead on two fronts to produce *two* new models. Standard-Triumph, at the time, was renowned for this. The result was a triumph (sorry!) for Standard-Triumph's 'can-do' engineering skills. First of all, it was decided that the new sports coupé, to be called GT6, and the up-graded Vitesse, should both be given the full 1998cc version of the six-cylinder, which was physically the same size as the 1.6-litre version already used in the Vitesse and in the 'Spitfire GT' prototype. This was already being fitted in the Triumph 2000 saloons and estate cars, and it was also proposed to fit another version of it into a revised TR sports car in due course, so it all made a great deal of planning and marketing sense. For this new application it was given a power boost to 95bhp (5bhp more than the tune used by the 2000 family car), which quite coincidentally – or was it? – made it *exactly* as powerful as the newly announced MGB GT.

Next, the existing Vitesse's gearbox was discarded in favour of a new four-speed all-synchromesh design – this, in fact, being the most significant item of capital investment in the build-up towards the two new cars. This gearbox was actually used in the 'works' competition Spitfires from late 1964, and proved to be almost completely reliable in that form. A new back axle design, with more torque capacity, was also prepared, meaning that all the obvious weak points of the original Spitfire/Vitesse drive line had been eliminated.

Very little change was needed to the Spitfire chassis and body to accommodate the new running gear. The bonnet panel needed re-shaping to clear the longer six-cylinder engine, so it was given an attractive 'power-bulge' which just happened to be rather reminiscent of that found in the contemporary and charismatic Jaguar E-type. The new long sloping roof panel (tooled, and pressed in quantity, by Pressed Steel, though the rest of the shell, and final assembly, were concentrated at Forward Radiator in Birmingham, alongside the existing Spitfire facility) actually fitted to the existing Spitfire windscreen surround, the rear wing, and the rear lower panel pressings.

In its original form the GT6 looked, and even sounded, rather like a miniature version of the Jaguar E-type coupé, but the GT6 production car exhaust systems were made slightly quieter and more refined than those of the prototypes, and some sportiness was lost. If only more work had gone into the improvement of the handling, the new car might have been an even more attractive proposition.

By the time the GT6 went on sale, the Triumph's swing-axle rear suspension was already well known to the pundits. In the wet, and in sudden lift-off-throttle conditions on corners, the rear wheels could tuck under rather alarmingly, thus pitching the cars into a sudden and sometimes terminal oversteering slide. There is nothing unique in this. Try out a similar maneouvre in the legendary Porsche 911 and observe the result . . . Triumph's engineers knew all about the problem, and had a 'fix' ready and waiting, but they were repeatedly assured

This was the first time a 2-litre six-cylinder engine was mated to the Herald family chassis frame. Originally 6003VC was a Vitesse 1600, but for the Spa-Sofia-Liège rally of 1963 Triumph's works competitions department built a 2-litre 'special', complete with TR4 gearbox and prototype TR4A rear axle. Following an engine fire the car was burnt out in Jugoslavia.

The GT6 was an amalgam of Spitfire and Vitesse engineering, with its own unique coupé body. This 1971 Mk3 was destined for sale in the USA.

34
6003 VC

by their sales force, especially in the USA where most of the sports cars were being sold, that the customers were not unhappy about it at all. They also listened to the sales force when they demanded a softer ride for the GT6 than the existing Spitfire – and gave it to them.

Thus it was that the GT6 was finally put on sale in the autumn of 1966 with a rather soft 'boulevard' ride, and with handling no better than any of the other separate-chassis Triumphs. Because of this, and maybe because they thought that their previous criticisms had been ignored, the specialist press were very critical of the first GT6s, and gave them something of a hard time in the road tests.

It would be two years before a revised and much improved version of the GT6 was made ready, and by that time the reputation

Bob Tullius of the USA was an established Triumph racing driver before he attracted Quaker State sponsorship and founded Group 44 for the 1970s. At this time he ran TR6s and GT6s – in the 1980s, of course, he ran the XJR5 and XJR7 IMSA Jaguars.

and image was cast in concrete. Nothing that Triumph (or the new owners, British Leyland) could do would alter this. For the next five years the GT6 was neglected by the public, and rather put on one side by the factory. It was no wonder that the arrival of more and yet more legislation from the USA was the final straw. To comply with it would have taken thousands of man hours and a great deal of investment. The GT6 was allowed to wither on the vine, and it finally died off before the end of 1973. 40,926 GT6s were produced, of which 22,658 were sold in the USA.

The 1960s was the decade in which Standard-Triumph's reputation for building attractive sports cars flourished, but in the 1970s that reputation gradually but inexorably died away. In 1970 the Spitfires and GT6s were much better than they had been a few years earlier. However, by the late 1970s, not only had they not been replaced by all-new models, but the final examples were arguably not even as nice as those built earlier. Both cars, at least, died with dignity, but it was a great and lasting shame that British Leyland could never find the cash – or the will – to produce Spitfire and GT6 successors to take the marque aggressively into the 1980s.

PERFORMANCE FIGURES

	Spitfire Mk 1	Spitfire Mk 2	Spitfire Mk 3
Max speed	92mph	92mph	95mph
0-60mph	17.3secs	15.5secs	14.5secs
Standing ¼ mile	20.9secs	20.0secs	19.6secs
Overall fuel consumption	31mpg	30mpg	33mpg

	Spitfire Mk IV	Spitfire 1500
Max speed	90mph	100mph
0-60mph	16.2secs	13.2secs
Standing ¼ mile	20.6secs	19.1secs
Overall fuel consumption	32mpg	29mpg

	Spitfire Mk IV/1500 – USA-spec. 1973	Spitfire Mk IV USA-spec. 1971
Max speed	94mph	86mph
0-60mph	15.4secs	15.9secs
Standing ¼ mile	20.2secs	21.0secs
Overall Fuel consumption	25mpg (USA)/31 (Imp)	30mpg(USA)/37 (Imp)

	GT6 Mk 1	GT6 Mk 2	GT6 Mk 3
Max speed	106mph	107mph	112mph
0-60mph	12.0secs	10.0secs	10.1secs
Standing ¼ mile	18.5secs	17.3secs	17.4secs
Overall Fuel consumption	20mpg	25mpg	28mpg

Spitfire Mk 1, 2 and 3

The original Spitfire was launched at the Earls Court Motor Show in October 1962, an occasion which also saw the arrival of the Lotus Elan, the MGB, and the Austin-Healey 3000 convertible; it was a great year for British sports cars!

Clearly the new Triumph was a direct competitor for the established Austin-Healey Sprite/MG Midget twins. It was equally fast, just as sporting, and was in many ways better, but as customers were asked to pay a bit more for the Spitfire – which had wind-up windows where the Spridget did not – the equation was finely balanced. As the show opened the Spitfire (now remembered as the Mk 1) was priced at £730 in the UK, but purchase tax reductions announced days later reduced this price to £641 – just £42 more than the Midget. Would it sell? The response was immediate – within months Triumph knew it had a winner.

Although the Spitfire's backbone-style chassis frame and the curvaceous bodyshell styles were all new, almost all of the running gear was modified (or in some cases pure) Herald family car. Independent front suspension by coil springs and wishbones, independent rear suspension by a transverse leaf spring and swing axles, allied to rack-and-pinion steering offering an amazingly tight turning circle, were all lifted directly from the Herald; only the spring and damper settings had been changed. Brakes, by front disc and rear drum, were all familiar 'Herald parts-bin' components. Although the Herald's 1147cc engine was in twin carburettor tune, the gearbox and final drive were pure Herald; only the gear lever had been changed.

The chassis frame was a robust backbone design linking engine bay and front suspension to rear suspension. The prettily styled steel bodyshell had a completely detachable front end, and was bolted to the chassis on assembly; the soft top had 'build-it-yourself' hood sticks which tucked away around the petrol tank when not needed. Unlike the Herald, whose layout was rather different, much of the assembly's torsional and beam strength was hidden away in the sturdy sills.

For a year the Spitfire was sold in its original 'basic' guise, but from the autumn of 1963 a

Triumph's Spitfire show chassis of 1965, this clearly being a Mk 2. Note the backbone chassis, which had no chassis side members running down the insides of the sills. Compared with the Mk 1, this Mk 2 had a tubular exhaust manifold and a modified cooling system.

steel hardtop, wire-spoke wheels and a Laycock overdrive all became options; such items have retrospectively been fitted to many earlier Spitfires too. This added to the appeal of an already attractive little car.

The first important change came in March 1965, with the launch of the Mk 2 type. Basically this was like the original Spitfire, but it had a slightly more powerful engine – 67bhp instead of 63bhp – and changes to the trim and furnishings.

Two years later the Mk 2 gave way to the Mk 3, which was a more substantial up-date of the same basic specification. The Herald-based engine was enlarged to 1296cc, there was a new eight-port cylinder head and peak power went up to 75bhp. To match the extra performance, larger front disc calipers were fitted.

The other big change was to the bodywork and fittings. Not only did the Mk 3 get a proper fold-away soft-top, but there was a modified

The Mk 2's front corner, demonstrating the slightly modified cooling system of the latest car. Older cars may show some signs of creasing around the welded joints of the front chassis members. It's easy to see how this could happen.

All Mk 1, 2 and 3 models used the same basic type of swing-axle rear suspension, which had a very high roll centre. The best roadholding is achieved by making sure the wheels have a slight negative camber when in the laden state.

The smoothly rounded detachable hardtop was a popular option. It became available in the autumn of 1963, but could also be fitted to the early Mk 1 cars. Note the brightwork cappings over the welded panel joints along the peak of front and rear wings. On neglected cars, rust often starts in those areas.

The Mk 3's styling was changed by the bumper being raised to this 'bone in teeth' position. This was not done for aesthetic reasons, but to make the car comply with new USA safety legislation.

front bumper, raised to meet pending USA legislation, yet more detail improvements to the cockpit (including a sprung-spoke steering wheel), and negative-earth (instead of positive-earth) electrics. A completely revised facia style was built into US-market cars for 1969, but this was never adopted on UK-market examples.

Further improvements were made in the autumn of 1969, when the Mk 3 was given 4.5in rim steel wheels instead of the original 3.5in rims, and at the same time there was a package of cosmetic improvements to the cockpit area. The car then remained unchanged, until replaced by the Mk IV at the end of 1970.

Engine

The Spitfire's engine was probably the most successful piece of hardware ever produced by Standard-Triumph. Designed in the early 1950s, it was unveiled (in 803cc form) for the Standard Eight saloon of 1953. In the years which followed it became a very important 'building block' for use in new Standards and Triumphs, and it was also the inspiration for the straight six-cylinder engine used in the GT6 and many other models. Before finally dropping out of production 27 years later, it had been used in cars as different as the Standard 10hp van, the Spitfire, the Triumph Herald, the Standard Atlas Van, the front-wheel-drive Triumph 1300 and 1500 saloons, and the West German Amphicar. Nor should we forget that it also found a home in the late-model MG Midget of the 1970s.

The 'SC' engine (SC = Small Car) of the 1950s was at once a simple, but up-to-date, design, with overhead valve gear, a chain-driven camshaft, and a three-bearing crankshaft. Although special aluminium cylinder heads were used on the 'works' Spitfire competition cars of 1964-1966, every production unit had simple and rugged cast-iron cylinder block and head castings.

Early types, as used in the Standard Eights, Tens, and original Heralds, were of 803cc or 948cc, but from 1961 the 'standard' size became 1147cc. This was achieved by re-shuffling the positions of the cylinder bores in the block casting. At the same time a new crankshaft design was needed, which involved a reduction in the width of the big end bearings (though lead-indium bearing materials were specified).

In Spitfire Mk 1 guise, the engine produced 63bhp, while for the Spitfire Mk 2 this was raised to 67bhp; the latter engine had a fabricated tubular exhaust manifold. In each case the familiar 'six-port' cylinder head was used – which means that there were individual exhaust ports, but siamesed inlet ports which brought some restriction in breathing

Spitfire and GT6 line-up: nearest the camera are (left) a 1500 and (right) Roger Rowley's Mk 3, behind them (left) a Mk 1 GT6 and (right) a Mk 3 GT6.

Study in tails: Spitfire Mk 3, 1500, Mk 1 GT6 and Mk 3 GT6.

Roger Rowley's Spitfire Mk 3 has won many concours awards in recent years. The Mk 3 was introduced in the spring of 1967 and featured a raised front bumper.

Interior of the Mk 3 Spitfire, distinguished from its predecessors by having a sprung-spoke steering wheel and a wood veneer instrument panel.

Engine bay of the Mk 3, with the 'new' eight-port cylinder head.

The early Mk 1 engine bay showing the radiator header tank above and inside the SU carburettors. Fibre-board side curtains were soon added to each side of the engine bay to protect it from water and filth thrown up by the wheels. This was a left-hand drive car – note that the battery and master cylinder locations swopped sides to make way for the re-arranged steering between left-hand and right-hand drive.

capabilities. That was why the 'works' competition cars were given a new 'eight-port' cylinder head, for this transformed the peak power potential of the unit.

For the Spitfire Mk 3 there were more substantial changes. The front-wheel-drive Triumph 1300 saloon had been launched at the end of 1965, with a much-modified engine, and the latest Spitfire drew on this as its base. Not only was the engine enlarged to 1296cc (with a wider cylinder bore), and had a modified cylinder block, but it used a modified version of the 'eight port' head which had been introduced for the 1300, and a cast exhaust manifold. Later in the 1300 family car's life the more powerful twin-carb 1300TC was announced, this car sharing almost the same engine as the Spitfire Mk 3.

In every case, and for every market, these Spitfires used twin SU carburettors, though USA-market Mk 3s had various extra 'hang-on' components to keep the exhaust emissions in check.

By late 1980s standards these Spitfire engines are extremely simple units, for there are no hideously 'high-tech' electronic components to frighten the DIY owner. Indeed, the most complex operation on a Spitfire engine is re-balancing and setting up the constant-vacuum SU carburettors after a rebuild (or if they have gone out of adjustment with wear and old age.)

In general, these engines – 1147cc or 1296cc – are renowned for their longevity, with many completing more than 100,000 before needing major attention. The engine was never an ultra-quiet unit, for it sounded rather 'thrashy' when revved hard. Extra-noisy valve gear and heavy oil consumption tell their own story, but tired engines can often be rejuvenated by a thorough top-end overhaul. With old engines look (or listen) for

stretched camshaft drive chains, worn crankshaft thrust washers and – almost certainly – SU carburettor installations which have gone out of adjustment. Although new head and block castings are extremely rare, every other part can be bought, to produce an 'as-new' engine for the *next* decade.

Gearbox and rear axle

The Spitfire's *basic* gearbox design, like its engine, was a component used in many other small Standard and Triumph cars over the years. Like the engine, it was new for the Standard Eight of 1953, modified for the Pennant of 1957, further modified for the Herald family of 1959, and refined further for Herald 1200 and Spitfire range. In close-ratio form it was also used in the Vitesse 1600 range.

By the time the Herald 1200 and Spitfire came along, the design featured a cast-iron gearbox casing, bolted to a cast-alloy clutch bellhousing. Previously, a one-piece alloy casing-bellhousing had been used – which means that older models cannot be 'robbed' to assist in rebuilds of more modern cars.

The first three Marks of Spitfire all shared the same gearbox, complete with high-mounted remote control linkage, and three-rail selection arrangements inside the top cover. There was no synchromesh on first gear, and since the basic design had originally been conceived to deal with no more than 33bhp in the Standard Ten of 1954, it had a hard time dealing with twice as much as this for the Spitfires.

Except in detail (such as the shape of the gear lever, and the lever knob itself) the Spitfire shared precisely the same box as Heralds of the period. The gearbox used in

Engine bays changed slightly from model to model. On this Mk 3 there were noticeable differences in carburation, air cleaners and related fittings compared with the Mk 2 which it replaced. In particular, the engine's cylinder head had four inlet ports, and completely different manifolds to suit.

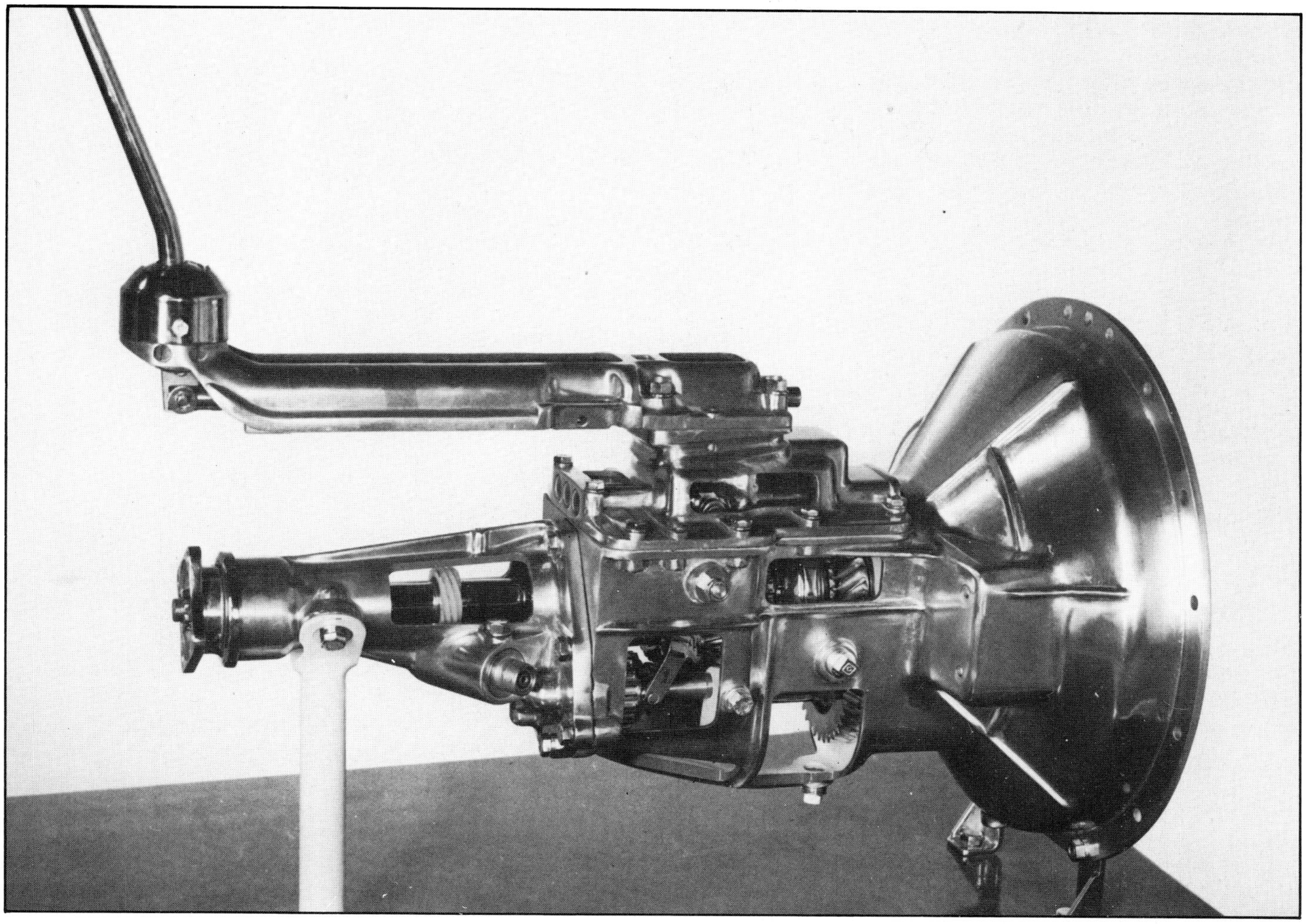

The 1959 Herald's gearbox featured a light-alloy combined gear casing and bell housing. In the interests of cost saving and sound-deadening, the casing became a separate iron casting by the time the Spitfire was launched. The remote control change was used on all GT6s, and all Spitfires built up to the end of 1974.

Mark IVs and 1500s was different in almost every way.

Mk 1s had a 6.25in diameter single dry plate clutch, but a diaphragm spring clutch was specified for the Mk 2 and later models.

The smallest Laycock overdrive (the D-Type) became optional from late 1963. The same 'kit' was also optional on the Vitesse 1600, though never officially on any of the Heralds; in modified form it was also fitted to Vitesse 2-litre/GT6 models which, of course, had a different gearbox.

By comparison, say, with the BMC-gearboxed Sprite/Midget cars, these Spitfires never had such a sweet feel to the gearbox; this was mainly due to the articulation of the lever, and to the rather wide gate. The synchromesh on this 'box (remember it was quite small and not really intended to look after 65-75bhp!) soon wears out on sportily-driven Spitfires, and it can also get somewhat noisy. Nylon bushes in the selector mechanisms eventually wear, and the linkage becomes sloppy. There was an unfortunate and built-in tendency for the gear lever to jump out of mesh when the car was using high torque in reverse gear. The only cure for a 'box which habitually jumps out of reverse is to treat it to a full rebuild. Fortunately, if the casing is still in good condition, this is still possible as several specialists provide an exchange 'box service.

Overdrives, where fitted, are desirable because they add at least 10 per cent (and sometimes more) to fuel economy, and make the Spitfire a very relaxed car to drive on late-1980s style roads. As with every Laycock unit, these suffer much more from dirty oil (and clogged internal passages) and failed exterior electrical solenoids, than they do from severe wear. However, rebuilt units are available, on an exchange basis, from several specialists.

Incidentally, although it is possible to add an overdrive to a non-overdrive car, this is a rather complex business, involving a complete gearbox strip-out, wiring changes, and many other related details. I considered it at one time on my own Spitfire Mk 1, but saw the implications and gave in gracefully!

The back axle of the earlier Spitfire was not its strongest point. As with the gearbox, the size of the crown wheel and pinion, together with the differential itself, had been conceived in the early 1950s for the Standard Eight/Ten models. The Spitfire had to deal with much more power and torque than these cars, and breakages of output shafts and components were by no means unknown.

The axle, however, was gradually improved over the years; when the 2-litre engined GT6s and Vitesses were introduced the axle was made considerably more robust. Because it is possible to mount a later type of axle into an early chassis, it is now quite possible that an earlier model has been given a more robust assembly (perhaps even with a non-standard final drive ratio) than it originally had.

Parts expert John Kipping has kindly provided the following information, which lists the strongest and the not-so-strong back axle assemblies. The chart below lists Herald/ Vitesse cars as well as Spitfires and GT6s. He does *not* recommend using Herald 948cc or pre-1962 Herald 1200 axles, as these were not strong enough.

The three important variables are:

1. Flanges (connecting the differential to the half-shaft or the propeller shaft) – ½in or 9/16in spanner bolts;
2. Pinion bearings – S for Small, L for Large;
3. Output shafts – S for Small (and rather weak), L for Large and really unbreakable.

No adverse effects occur when a later-model four-stud casing for holding down the transverse leaf spring is used in an earlier car originally designed for a six-stud casing.

Suspension, steering and brakes

The Standard Eight/Ten/Pennant series of the 1950s established a basic coil spring front suspension layout which was carried over to the Herald and beyond. However, since the Herald's suspension was different in almost every detail, it is best to consider it as a new design whose production life began in 1959. Not only did the Herald have a simple, rugged, easy-to-repair layout, using combined coil spring/telescopic damper units and wide-based double wishbones, but it was allied to rack and pinion steering. Even if it had been

MODEL	UK Axle Ratio	Prop. Flange	Pinion Bearing	Output Shaft	Output Flange
Spitfire 1, 2, 3 (to Chassis FC120000)	4.11:1	S	S	S	S
Spitfire 3 (from FC120001)	4.11:1	S	S	L	S
Spitfire IV *	3.89:1	L	L	L	L
Spitfire 1500 *	3.63:1	L	L	L	L
GT6 *	3.27:1	L	L	L	L
GT6 *	3.89:1	L	L	L	L
Herald 1200 (to Chassis GA237600)	4.11:1	S	S	S	S
(from Chassis GA237601 and all 13/60s)	4.11:1	S	S	L	S
Vitesse 1600	4.11:1	S	S	S	L
Vitesse 2-litre *	3.89:1	L	L	L	L

* All these units are completely interchangeable, without any flanges needing to be changed. Unless you are a stickler for period authenticity, it is clearly best to fit the strongest axle which gives the correct ratio. Stocks of parts are good, and many Triumph specialists sell exchange final drive units.

The wishone front suspension of Spitfires and GT6s tends to be neglected as they get older. In particular the trunnions at the bottom of the vertical links run out of lubricant and their threads wear badly. After a front end accident wishbones or their chassis mountings tend to be distorted. This picture shows a new and correctly set up assembly.

conventional it would have been an excellent suspension, but the fact that it also provided a remarkably compact turning circle (25ft between kerbs – the tightest in the world at the time) made it exceptional. Naturally it was also used on the six-cylinder Vitesse and on the Spitfire.

The front suspension of the Spitfires Mk 1, 2 and 3 was the same as that of the Herald family car, except for the use of different spring and damper settings. Because the Spitfire's wheelbase was shorter than that of the Herald, its turning circle was a miraculous 24ft, which I believe has never been beaten by any other car, not even the London taxicab.

An old and neglected Spitfire can be a rattly, jolty, imprecise, machine to drive, and much of the blame falls on worn out front suspension assemblies. The wear pattern is well established. Treat the wishbone, anti-roll bar and steering rack mounting bushes (rubber or nylon, depending on the particular point) as consumable – all can be obtained, and renewal is simple enough for a competent mechanic. Rack mountings may become impregnated with oil and allow the rack to move around a little, which destroys the precision of the steering and can feel rather spooky. Most of all, however, be prepared to change the vertical link (lower) trunnions at regular intervals, perhaps every 25,000 miles or so. The reason for this is that the trunnion threads *must* be lubricated at regular intervals, something which has often been neglected,

Spitfires 1, 2 and 3 could often suffer from sudden oversteer on corners. This was caused by the 'jacking-up' of the rear suspension, as demonstrated graphically in this racing circuit demonstration shot.

especially if a previous owner did not have a workshop manual (or did not care too much about preventative maintenance).

If such a Spitfire has had an accident it is likely that the front suspension mountings will have been 'tweaked', or the wishbones may have become distorted. Always be sure that the suspension/steering geometry is nominally correct, and don't hesitate to replace any parts if you find errors.

The Herald of 1959 was the first Standard-Triumph model to use independent rear suspension. This featured a transverse leaf spring clamped to the top of the final drive casing and fixed-length driveshafts with inner, but no outer, universal joints, and provided cheap and cheerful swing-axle geometry. Fore and aft location was by radius arms linking the hubs to the chassis frame, and telescopic dampers were mounted immediately behind the line of the spring. Except for the 'set' and the rate of the transverse spring, and the setting of the dampers, the Spitfire and Vitesse's rear suspension was identical; in the case of the Spitfire, the radius arms attached directly to the bodyshell (behind the seats) rather than to a chassis outrigger.

The roll centre of this layout was very high, there was also considerable wheel camber change between full bump and full rebound, and the wheels and tyres used were narrow, so the cars tended to develop abrupt oversteering tendencies under hard cornering, particularly if the throttle was released in mid-maneouvre. Very few people ever sought to justify retaining this layout for so long – the few, however, included the directors and senior engineers of Standard-Triumph.

There was a lot of tolerance in rear spring height settings. Spitfires with static negative rear wheel camber handled much better than those with static positive camber; a skittish car can often be cured by the fitment of a different (or de-cambered) rear spring.

Rear suspensions, like the front end, often become rattly. This is usually due to worn bushes in the radius arms or damper mountings or where the vertical link is

This Spitfire Mk 1 has the right degree of rear wheel camber, which will assume a slightly more negative aspect when the driver climbs aboard. The sharp-edged panel seams under the sills and behind the rear wheels are obvious places to inspect when searching for corrosion.

connected to the transverse leaf spring. Be prepared to change bushes every 25,000 miles or so – more frequently if the car has a hard life. Be sure that the forward end of the radius arm is still secure in its fixing to the bodyshell.

If 'your' car has a rear spring which gives the car positive static wheel camber, consider buying a replacement with a lower setting and less camber – the effect on roadholding is certainly worthwhile.

The original Spitfire used the same pressed-steel wheels as the Herald 1200, with 3.5in rims. The pity of it all was that the Courier van was already on sale with similar wheels having 4.5in wide rims, but for cost reasons these were not used at first. Differently-styled wheels, with 4.5in rims, were adopted in the autumn of 1969.

From late 1963, the option was a set of smart centre-lock wire-spoke wheels, which became very popular. These always had 4.5in rims, and were the same as those used on GT6s. They were not, however, as robust or as light as the pressed steel wheels, which explains why they were never used by the 'works' competition departments.

In later years many Spitfires have been treated to a set of ultra-wide wheels of proprietary manufacture, but these needed fatter-section tyres to match. They did little for the car's handling, nothing for the looks, and gave the wheel bearings a hard time.

Old wire wheels tend to have loose spokes, and are not as rigid as steel disc wheels. The 'running a screwdriver round the spokes' test (to listen for 'pinging' or 'clunking' noises) works well. There is still wire-wheel rebuilding expertise in the UK, but this does not come cheap. In old age the centre splines wear rather badly, which allows radial movement between wheels and hubs; this rapidly accelerates the wear.

The original Heralds had drum brakes all round, but from autumn 1961 9in front discs became optional. The Vitesse 1600 always used the same front brakes, but had 8in instead of 7in rear drums. For the Spitfire, therefore, is was easy to specify the 'optional Herald' installation – 9in diameter front discs, and 7in diameter rear drums – for no extra development was needed. These basic dimensions were retained throughout the life of the Spitfire, from 1962 to 1980; at no time were the 8in diameter drums of the Vitesse fitted to a production car. (There was ample space inside the 13in wheels for larger brakes to be fitted. The 'works' race and rally cars always used them – TR4-style 9in drums being used at the rear for Le Mans in 1964.)

There are no out of the ordinary problems with these Girling installations, and a long-travelling handbrake lever is normal. The handbrake cables, however, have a tendency to rust up, due to lack of use or neglect during storage.

Chassis

Too many people glibly suggest that a Spitfire chassis frame was only a short wheelbase version of that of the Herald. It was not – except for a few common brackets and mountings the two frames were completely different. The Herald's frame relied on backbone *and* side-members to give beam and torsional strength; the Spitfire's frame was a pure backbone design, with no outer side members. The main backbone had box-section members, and there were vestigial outriggers abeam the clutch bellhousing to give stiffness under the footwells. At the rear there were no outriggers, the suspension radius arms picking up directly to the bodyshell itself. Except in detail, the Spitfire's frame was the same as that fitted to the original GT6, except for different engine and radiator mountings. Although the chassis is a robust item, it can, and does, corrode badly over the years, and at the front it is also vulnerable to accident damage – the usual evidence is that a front 'corner' has been straightened, or a new section has clearly been welded in. Sometimes this will not have been done exactly, and the result will be a badly fitting bonnet (which is hinged at the front of the chassis).

Body support outriggers sometimes rust away badly where they join the chassis. Replacements are available, but they must be positioned *very* accurately at the welding-up stage. Rusting and distortion around front suspension mountings should always be rectified during a ground-up rebuild.

Body

The Spitfire's body was different in every way from that of the Herald/Vitesse family; it is important to stress this. The Herald family of bodies was built up from bolt-together sections, and in several different guises, contributing very little to the overall strength of the body/chassis structure. Except for its bonnet/wings assembly, the Spitfire shell was welded up on assembly and the box-section sills added significantly to the overall rigidity of the structure.

Except for cosmetic differences, the Spitfires 1, 2 and 3 all used the same sheet metal bodywork. An obvious improvement for the Mk 3 was that a more modern, fold-back, soft-top assembly was fitted; earlier types had tubes just behind the door openings into which simple hood sticks had to be inserted. When stowed, these were slotted into place at each side of the fuel tank, from the boot area.

It is important to realize that the windscreen could be removed, and that the heelboards (behind the seats) were used to pick-up the rear suspension radius arms. For obvious reasons (suspension geometry and structural integrity) these have to be watched as corrosion sets in with old age.

The bonnet/wings sub-section was hinged in the nose from brackets mounted on the front of the chassis, and was secured by Herald-type over-centre clips at each side of the car, close to the door hinges. The steel hardtop, if fitted, was fixed to existing body and windscreen rail mountings; although not light, fitting and removal was simple.

The hole in the outer sill under the door of this early Spitfire might hide even more ghastly horrors inside. Unhappily, many Spitfires which have been bodged up for sale have had filler skimmed over this area.

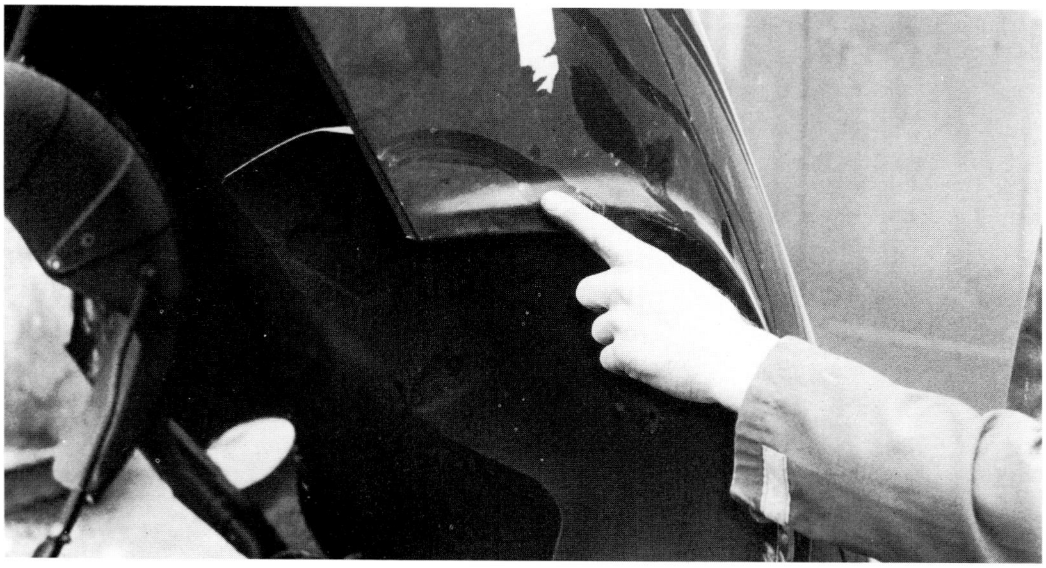

As they get older, all Spitfires (this is a 1500) begin to rust around the edges of the front wheel arch cut-outs. While you are looking in this area, take time to inspect the welded seam between inner and outer wheel arches, the line of which can be seen under the pointing hand.

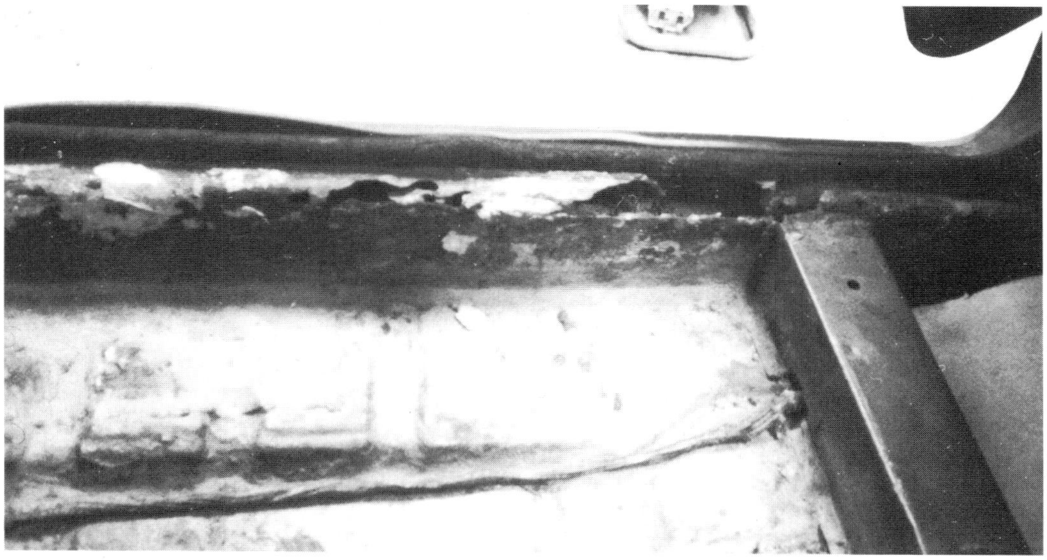

Extensive rot in an inner sill.

A 'home-made' inner sill, crudely welded together.

far left

The lower corner of this windscreen frame has rotted right through and been patched with glassfibre.

left

More 'repairs' – here a liberally holed footwell has been covered with a loose sheet of very thin metal.

bottom left

Filler has evidently been used here to conceal holes in the sill.

below

Little remains of the bodywork below the rear bumper on this Mk 3.

bottom

Spitfire doors rust in the usual places; the bottom has all but fallen out of this one, and there is rust all round the crimped-over edge of the door skin.

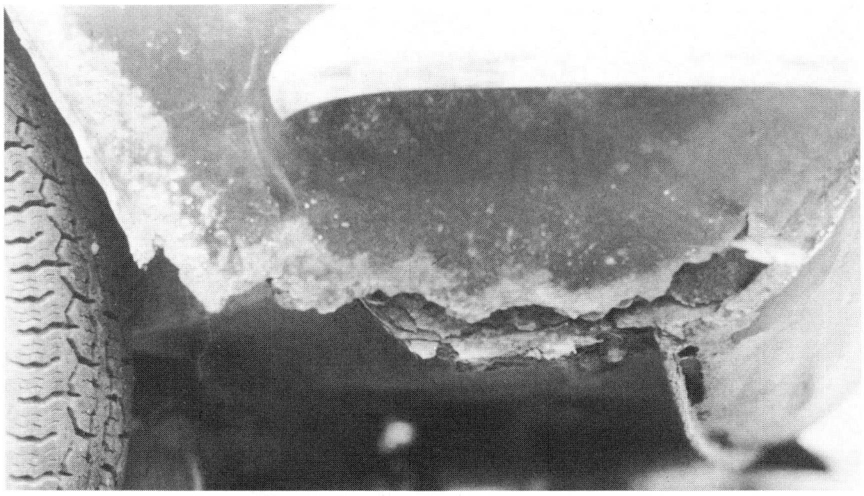

right
Extreme horrors ahead of the front wheel . . .

far right
. . . extending well into the inner wing.

below
Rust around and below this headlamp and alongside the finishing strip on the wing will eventually turn into holes . . .

below right
. . . as will the rust around this tail lamp and the finishing strip above

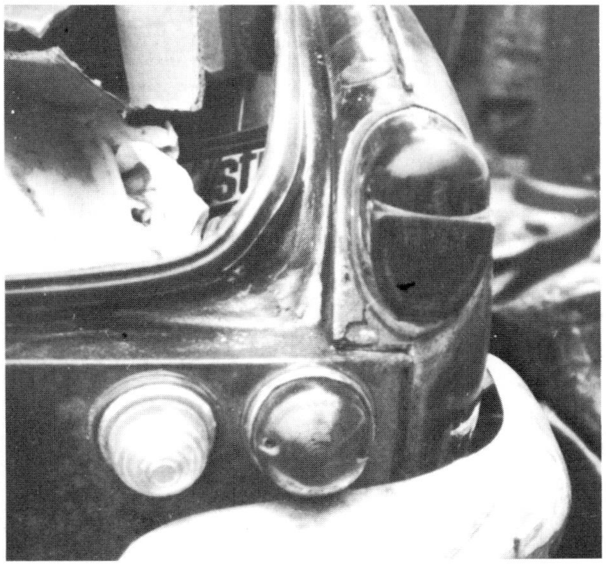

This is a Mk IV's wheel arch but its ailments are common to all marks of Spitfire.

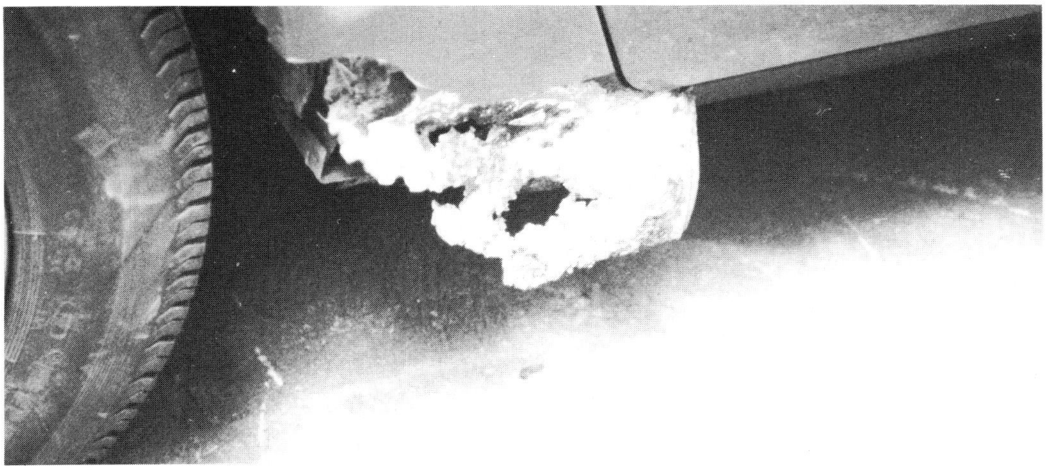

This Spitfire had been very recently botched up and resprayed: but some expert probing ahead of the rear wheel reveals the awful truth. This car is now being broken up.

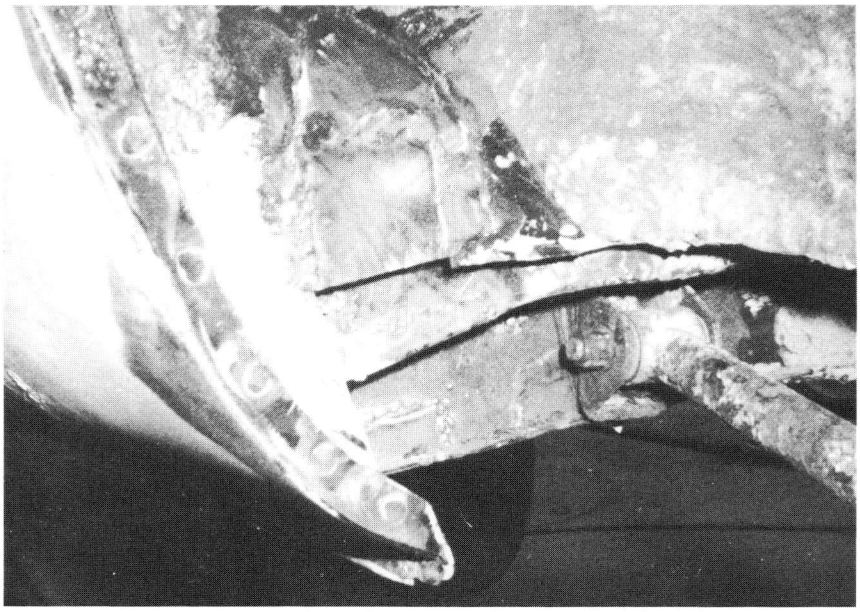

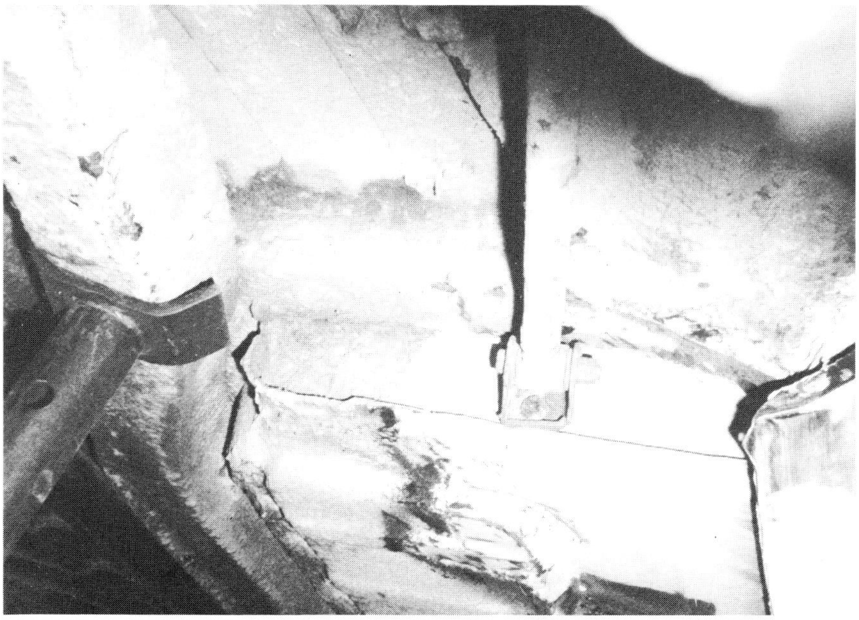

Very important: the radius arm pick-up points and surrounding area. These two pictures show repairs at different stages on both sides of the same car. This one is actually a Mk IV.

Inevitably, as the Spitfire body grows old, it rusts. When inspecting a car, start by checking for rust around the side-indicator cut-outs, and close to the headlamps, around the bonnet latch piercings, and along the seams linking bonnet to wing panels. Badly neglected cars often have inner wheel arches coming adrift from the rest of the bonnet assembly.

In the main shell, look for rusty (or even holed) floor panels, and for badly deteriorated sills (especially inner sills). In the boot there may be bad rusting in the floor, around the inner wheel arch joints, and above the axle. The most vital spot to look for is the pick-up point for the rear suspension radius arms. Doors sometimes rust badly at their bottom edges, and there may be rust on the scuttle near the base of the screen or heater air intakes. Rear wings suffer in front of the wheel arch cut-outs, along the edges of the arches, and below the rear bumpers.

A rusting Spitfire is not a pretty sight . . .

Interior

Over the eight years covered by this section, Spitfire interiors progressed from being spartan to being simple but completely furnished. Except for late-model US-market Mk 3s, there were no basic changes to the facia layout during two model changes. It featured a central instrument panel and a scuttle brace down to the chassis on each side of the gearbox tunnel. The brace had a definite purpose – for stiffening the assembly and stopping the scuttle from shaking – as all who have tried to run without it soon discovered! The USA-market models mentioned above used a different facia layout, with speedometer and rev-counter mounted behind the steering wheel, ahead of the driver's eyes.

By comparison with later types, the Mk 1 had a rather starkly trimmed cockpit, with a gret deal of painted metal on show. Mk 2s had all this covered over by trim material, and Mk 3s were plusher still. Moulded rubber mat floor coverings are no longer available – why not fit carpets instead?

The Mk 2 had a more completely trimmed cockpit than the Mk 1. Some owners have fitted Mk 2 carpets (or replicas of them) to earlier types.

Mk 1s had moulded rubber 'carpets' which soon wore out, and almost every survivor has retrospectively been improved by the fitment of the moulded carpets used on Mk 2s and Mk 3s. There was too much exposed painted metal in the cockpit of Mk 1s, which was covered by trim material on later types.

Seats were simple buckets, with no adjustment for backrest angles. Steering wheels were shared with other Triumph models – Mk 1 and 2 used Herald/Vitesse types, while the original Mk 3 wheel was like that of the TR4. The final Mk 3s used the same wheel as the Mk IV, flat alloy spoked, like that of the GT6 Mk 3 and the later TR6s. Steering columns, in any case, were from the Herald/Vitesse range.

On a Mk 1, it isn't worth trying to patch up the rubber floor mats (if still fitted). Forget all about originality – throw them out and fit Mk 2 or Mk 3 floor carpets instead. To change floor coverings, the scuttle brace has to be removed – don't worry, the facia won't collapse when you've done this!

Older cars often have worn or torn trim panels and seat covers. If you are a proud owner, it's nice to know that new 'replica' panels are now available; it takes a real expert to tell a replica from an original – and since you can't get originals anyway . . .

Each type of Spitfire was noticeably different from its ancestors. This was the cockpit of a Mk 3 – note the spring-spoke steering wheel and the wood veneer on the instrument panel.

This old and rather scruffy Mk 2 shows how easy it is for the boot area to get into a bad state. The jack is no longer in its specified bracketry, and the panels inside the boot are well and truly scuffed.

Restoration

Although there are plenty of Spitfires left in the world the supply of Mk 1, 2 and 3 cars is inevitably more restricted than that of the Mk IV/1500 're-skinned' models. However, as there is now a healthy parts re-supply industry and several well-respected specialists in the marque, restoration of a thoroughly scruffy example is always feasible. It is not *quite* possible to build up an authentic Spitfire entirely from components, but as long as the 'basket case' you buy has an undistorted chassis and the basis of a recoverable bodyshell the rest is only a matter of time, patience and – I'm afraid – money.

One of the joys of Spitfire restoration, if you are determined to do it yourself, is that the car is really no more than a monstrously large (and exciting) construction kit or Meccano set. On the assumption that you have enough space available to store everything, the bonnet and front wings can be removed to give perfect access to the engine bay and front suspension, while the entire bodyshell can be hoisted off the chassis if you locate (and loosen – rusted-up bolts can be remarkably tenacious after a number of undisturbed years!) all the fixings.

There are few high-tech installations (such as Hydrolastic suspension, electronics, or complex facia installations) to worry you. It is really only a matter of labelling everything when you take it off the car, and not losing it.

Not only are all the relevant Parts Manuals and Workshop Manuals freely available (see Clubs and Specialists), but a number of highly competent and extremely knowledgeable Herald/Vitesse/Spitfire/GT6 specialists have been established during the 1980s.

Engines, gearboxes, overdrives and rear axles can all be purchased as reconditioned 'black box' items. This means that if you simply cannot face a rebuild yourself (and if finance permits) you can go along to your chosen specialist and do a part-exchange deal.

The proper inspection of a Spitfire, prior to restoration, should really start by removing the body from the chassis, and as soon as possible after this the engine and transmission assemblies should then be removed.

Triumph restoration expert John Woolley, of Gloucestershire, advised me about the correct procedure to be followed, and I am extremely grateful for his help.

The complete bonnet assembly should be removed first, and then the bodyshell can be unbolted from the frame – there are 10 simple fixings. The fixings one might miss are the two at the back of the chassis above the differential and the two (usually hidden beneath underseal) under the seat squab. Unlike the Herald/Vitesse cars, there are no special sound and vibration insulating washers.

You shouldn't have too much trouble loosening most of the bolts except for the front pair, which are on the front outrigger under your feet in the footwells. These usually have to be sawn off, and new ones used in reassembly.

The first job is to get the chassis frame stripped of all paint and other coverings. Take it all the way back to bare metal, preferably using a high-pressure cleaner, then have it shotblasted. At this point any dodgy sections or rusty joints will be exposed.

An alternative is to remove the paint chemically, but this tends to take longer and can leave more mess to be cleaned up. By scraping and using chemical means, of course, any DIY enthusiast can strip down his own frame.

As soon as the frame is clean you should inspect it for bodged-up accident damage, or evidence of any sections which are not 'square' and sound. Repairs, in fact, can even be done before shotblasting is tackled. Plating, incidentally, is perfectly acceptable – as any 'Vintage' type will tell you!

Certain sections of the chassis are not available and, of course, complete new chassis are no longer available either. However, in planning the refurbishment of the frame, be ruthless in discarding any badly rusted sections which you can be sure are in stock. Don't even start wielding the big hammer and cold chisel before you have checked on availability. At the same time, spend a great deal of time making sure that the frame is still in correct alignment, and not 'racked' or twisted due to previous accident damage.

Aside from giving you peace of mind, a frame that is true can be used as a jig for refitting the body, so if you have trouble fitting the body or the bonnet then there is something wrong with the sheet metal.

At the front of the frame, the outriggers holding the bonnet pivot points have often been distorted or may have rusted. These are not large pressings but their correct alignment, along with that of the front crossmember, is vital to the correct fitting of the bonnet. The correct position of these pivot points 'in space' is shown on frame assembly drawings reproduced in the appropriate service manual.

The suspension towers (which are common with those fitted to the Herald/Vitesse and GT6 models) may have been damaged in a collision but it is unusual to find them rusting. They are bolt-on items.

The front outriggers are the most prone to rusting, usually adjacent to the chassis, but the centre outriggers rarely seem to give trouble. There is very little to worry about at the rear, unless you encounter some rust in the top sections above the back axle. The main chassis below that area is usually sound, even on a very old car. Problem areas have to be plated, as new sections are not available.

Once the frame has been reconstructed it should be shotblasted and then powder-coated. You can have what colour you like – many customers like their 'new' chassis to be basically the same colour as that of the bodyshell itself. To make a truly professional job, however, you might also like to have the bare frame zinc-coated before applying the powder coat – such a frame job will last for ten to twenty years if it is also treated to a Waxoyl (or similar) coating.

If you are not a stickler for restoration accuracy, consider strengthening the front outrigger/pivot points for the bonnet mountings – concours judges might not like to see extra gusseting in that area, but it is well worth doing.

The bodyshell can be treated in two parts: the bonnet assembly and the main structure.

Complete bonnets are no longer available, but outer wing sections and both pressings of the inner wheelarch can be found. The main centre pressing is not available. It is very easy to distort the assembly, particularly the wings, when rebuilding a bonnet assembly, but the problem is minimized if the tubular frame is left attached. Frankly John Woolley suggests that this particular aspect of body renovation should be left to a Triumph specialist.

Now for advice on how to get the bonnet assembly, or the main body structure for that matter, down to the most basic workable condition. Let's start with a negative: don't start taking the sills off the shell unless the body is supported by and bolted down to the chassis or it will buckle. The sills in a Spitfire are (or should be) extremely strong, and are structurally important.

To obviate this buckling, specialists have frames or jigs which fit across the door frame and hold the front and rear pillars apart, but for the private owner the chassis frame is an excellent jig in itself.

If you know that certain panels or sections are to be discarded, get rid of them at once, before the cleaning down process begins. Although most panels are available, one area that does rust very badly – where the front frame outrigger bolt comes through the floor, and the panel leading forward from it – is not. This will have to be plated.

It isn't wise to take all the body down to bare metal at the same time as a complete restoration will certainly take months and maybe even years. However, it is always wise to dismantle the shell as much as possible, as soon as possible – which is to say remove the doors, the boot lid, and the windscreen frame. Tackle repair and restoration one area at a time, and as soon as it is done, protect it immediately, perhaps not with top coat but at least with the usual zinc coating, undercoating and related treatments.

If the floor of the car is in bad condition this should be tackled first, then covered up so that you can sit or lie in it to tackle other sections. It needs to be as structurally strong as possible before the (usually necessary) sill replacement is tackled. Renewing floor panels, by the way, might distort the alignment of the tunnel, and since there is only minimal clearance between tunnel panel and propeller shaft this needs to be got right.

Next, tackle the sill replacement, something which can make a great deal of difference to the rigidity and the whole feel of the body/chassis assembly.

Although lower front edges and lower rear edges of the rear wings are available (to renew rusted patches), John Woolley recommends a complete wing change on all occasions. To remove an old wing it is necessary to remove lines of spot welds along the crown line of the wing, at the pillars, around the inner arches, and the tail panel. At the time of writing, such wing pressings are still ex-factory stocks, and need no fettling before fitment.

Surprisingly, the inner wing panel – the much-perforated straight panel inboard of the outer pressing – very rarely needs to be replaced, though there is sometimes corrosion at the joint with the boot floor. Arches, on the other hand, often have to be replaced.

below
A 'restored' Mk 2: note the excessive gap ahead of the door, the badly fitted outer sill, and the lower corner of the door actually overlapping the sill. Much work has gone into this car but it will have to be taken apart and started again.

bottom
The same Mk 2, showing the poor fit of the trailing edge of the passenger door.

Standard-Triumph completely re-skinned the Spitfire for the Mk IV and this style was continued, unchanged, for the 1500.

Spitfire 1500 on the move.

Unlike some rival sports cars, the Spitfire did not lose its good looks as facelifts were applied. The under-bumper spoiler seen on this 1500 was added from 1973.

Spitfire Mk IV and 1500 models had 1.3-litre and 1.5-litre engines respectively, both being of the same venerable Standard-Triumph 'SC' type. The trunking forward of the air cleaner drew fresh air from the front of the car.

The Spitfire 1500's nicely furnished cockpit. The gear lever knob on this car is non-standard.

far right
The driver's door of the Mk 2 doesn't fit snugly into the bodywork.

right
Much shiny red paint has been applied, but rust is breaking out around the door hinge mounts.

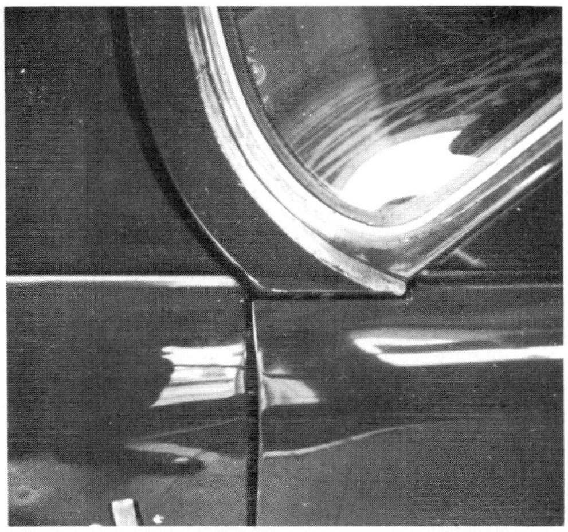

Gaping at the leading edge of the door, accompanied by poor fit of the bonnet section to the scuttle.

below
When restoring a Spitfire it is important to get the details right. On this Mk 3 note the position of the boot lid badging, and the fact that the silencer tailpipe was not chrome plated.

When all the patching, replacing of panels, welding and refurbishing are done, the body (if off the chassis at this point) should be bolted down and checked for alignment. Then it should be removed again. All the seamed and welded joints should then be treated with 3M Super Seam sealer, which stops water getting inside the joints and starting up the rusting process again.

Next apply two coats of a zinc primer, or something similar. John Woolley then recommends the use of a 'stone chip' compound, rather than any form of underseal, for the underside of the car. Then paint the car in the normal way.

The crunch point in reassembling a Spitfire comes when the body has gone back on the chassis and the bonnet assembly has been fitted. How is the fit of the bonnet around its rear edge where it lines up with the scuttle and the doors? Many Spitfires seem to have badly aligned bonnets, and there can only be two reasons for this – that the frame is bent, or that the outriggers supplying pivot points are in the wrong place. Both of these problems should have been rectified during restoration of the car, but it may still be necessary to make small changes to the pivot positions.

To summarize, meticulous attention to alignment is vital to the successful renovation of Spitfire bodies and chassis. For that reason, John Woolley recommends that the bodyshell should not be rebuilt while it is off the chassis frame, which provides invaluable support at every critical stage.

Maintenance

In almost every way the Spitfire is a conventional car, needing conventional maintenance. However, since it is a car which has already been forgotten by ex-Triumph (now Austin-Rover, or perhaps even Japanese-car) dealers you must be prepared to do your own maintenance and running repairs, or budget for the car to be returned to a specialist from time to time. A number of selected specialists are listed at the back of this book.

It follows that you should buy copies of the correct Parts Manual and the Workshop Manual (these are readily available as you will already know if you are a member of the Triumph Sports Six Club) and maintain the car accordingly. There are no short cuts, and no benefits, in ignoring the work scheduled. Modern oils and greases may be better than their 1960s equivalents, but the materials used in a Spitfire are just the same! Incidentally, with cars like these, it is *always* better to change the engine and transmission oils more frequently than the Manuals suggest, especially if they now only clock up limited annual mileages. This matters especially in a car with an overdrive fitted, for dirty oil in an overdrive inevitably leads to a breakdown. Don't forget that a few litres of fresh oil – the correct grade, mind you – is always a better investment than a gearbox/overdrive rebuild.

The fact is that more Spitfires have fallen into disuse and wreckage because of neglect than because they have prematurely worn out. The archetypal case here, of course, is that front suspension trunnions have sometimes not been greased for years, the result being that the threaded portions rapidly wear away and ruin the feel, and the precision, of the car's handling and steering.

Look under your Spitfire when it is up on a ramp, or (securely) on axle stands, and you will see that there are several ideal places for mud and road filth to gather encouraging corrosion to set in. Comprehensive underbody protection is advisable and so is an annual clean out, preferably by steam cleaning.

Cost guidelines

In recent years the 'classic' value of Spitfires has increased a little, but by comparison with the equivalent MG Midgets they are still cheaper. As the Spitfire is basically a more robust, longer-lasting car there is no logic in this difference, and it means that all Spitfire types offer remarkable value.

Let me start by reminding readers not to believe that optimistic asking prices seen in advertisements will actually be achieved by the greedy vendor. In the case of the Spitfire there are so many cars on the market at any one time that one can always shop around to find the right car at the right price.

Because these cars all have separate chassis, it is certainly worth buying a 'basket case' if the frame is sound. Expect to pay no more than £100/£200, less if the bodywork is in an awful state, or if, say, the bonnet/front end is missing. As I have pointed out, you *can* restore or replace almost anything, especially if you are not too worried about absolute authenticity.

Although Spitfire 1, 2 and 3 prices do not always match my judgement, I am convinced that the later models in this group are the Best Buys. The three extras most likely to jack up the price (or make haggling more difficult) are centre-lock wire wheels, an overdrive, and the detachable hardtop. At 1988 values, reckon to pay £100/£200 over the odds for a car with all those features.

In a nutshell

These are the simplest, oldest and – in some ways – the most appealing of the Spitfire/GT6 family. Because of their age, however, there are more scruffy ones than good ones on the market at any one time. Even before buying a Spitfire, come to terms with its character, and its potential flaws; you should always drive the car (or one like it) before signing the cheque. If you are new to the marque, and not familiar with the cars, I recommend taking along a Spitfire enthusiast to see (and try) the car you consider buying.

Always remember that a Spitfire is not at all the same sort of car as a Sprite/Midget, being more spacious, prettier and better-equipped, but perhaps not handling as well; is this the sort of sports car you want? Remember, though, that the Spitfire has by no means reached the peak of its 'classic' popularity; restoration will provide 'added value', which is an investment for the future.

Because the Spitfire has a separate chassis, and because the entire body front end can easily be unbolted from the car, it presents an easier restoration job than most other cars of a similar age. It is also a car which must not be neglected once the original restoration has been carried out; make it a rule that you will keep a close look at all the vulnerable areas, particularly the bodyshell, and the suspension components, and keep the car correctly serviced at all times.

right

Once the Spitfire became popular, specialists rushed to produce optional hardtops and different bonnet assemblies. This kit clearly drew on the Jaguar E-type for its inspiration.

SPECIFICATIONS

Engine
Construction
Main bearings
Bore/stroke
Capacity
Valves
Compression ratio
Fuel system
Power
Torque
Transmission
Top gear
Final drive
Brakes
Suspension, front
Suspension, rear
Steering
Wheels/tyres
Chassis
Body

DIMENSIONS

Length
Width
Height
Wheelbase
Unladen weight

Spitfire Mk 1	Spitfire Mk 2	Spitfire Mk 3
In-line four	In-line four	In-line four
Cast-iron block, cast-iron head	Cast-iron block, cast-iron head	Cast-iron block, cast-iron head
Three	Three	Three
69.3 x 76mm	69.3 x 76mm	73.7 x 76mm
1147cc	1147cc	1296cc
Pushrod ohv	Pushrod ohv	Pushrod ohv
9:1	9:1	9:1
Two SU Type HS2 carburettors	Two SU Type HS2 carburettors	Two SU Type HS2 carburettors
63bhp (net) at 5750rpm	67bhp (net) at 5750rpm	75bhp (net) at 6000rpm
67lb ft at 3500rpm	67lb ft at 3750rpm	75lb ft at 4000rpm
Four-speed manual, optional overdrive on top and third from 1963	Four-speed manual, optional overdrive on top and third	Four-speed manual, optional overdrive on top and third
15.7mph at 1000rpm in top, 19.1mph at 1000rpm in overdrive top	15.7mph at 1000rpm in top, 19.1mph at 1000mph in overdrive top	15.7mph at 1000rpm in top, 19.1mph at 1000rpm in overdrive top
Hypoid, 4.11:1 ratio	Hypoid, 4.11:1 ratio	Hypoid, 4.11:1 ratio
9in Girling front discs and 7 x 1.25in rear drums	9in Girling front discs and 7 x 1.25in rear drums	9in Girling front discs and 7 x 1.25in rear drums
Ind. by coil springs, wishbones and anti-roll bar, telescopic dampers	Ind. by coil springs, wishbones and anti-roll bar, telescopic dampers	Ind. by coil springs, wishbones and anti-roll bar, telescopic dampers
Ind. by transverse leaf spring, swing axles, telescopic dampers and radius arms	Ind. by transverse leaf spring, swing axles, telescopic dampers and radius arms	Ind. by transverse leaf spring, swing axles, telescopic dampers and radius arms
Rack and pinion	Rack and opinion	Rack and pinion
Steel disc wheels, 3.5in rims, or optional centre-lock wire spoke wheels, 4.5in rims. 5.20-13in tyres	Steel disc wheels, 3.5in rims, or optional centre-lock wire spoke wheels, 4.5in rims. 5.20-13in tyres	Steel disc wheels, 3.5in rims (4.5in rims for 1970), or optional centre-lock wire spoke wheels, 4.5in rims. 5.20-13in tyres
Pressed steel backbone style chassis frame, with box section side members	Pressed steel backbone style chassis frame, with box section side members	Pressed steel backbone style chassis frame, with box section side members
Pressed steel bodyshell, with two seats and two passenger doors. Open sports car, with optional detachable steel hardtop. Entire nose section lifted up for engine bay/suspension access	Pressed steel bodyshell, with two seats and two passenger doors. Open sports car, with optional detachable steel hardtop. Entire nose section lifted up for engine bay/suspension access	Pressed steel bodyshell, with two seats and two passenger doors. Open sports car, with optional detachable steel hardtop. Entire nose section lifted up for engine bay/suspension access
12ft 1in	12ft 1in	12ft 1in
4ft 9in	4ft 9in	4ft 9in
3ft 11.5in	3ft 11.5in	3ft 11.5in
6ft 11in	6ft 11in	6ft 11in
1568lb	1568lb	1568lb

Spitfire Mk IV and 1500

Until you look closely, the Spitfire Mk IV seems to be very similar to the Mk 3 which it replaced. But look again – there are substantial body changes, there is a new rear suspension, a new gearbox, and important changes to the engine. If Triumph's stylists had had *all* their own way, the Mk IV would have looked a lot different, too.

The Mk IV, in fact, was a complete re-packaging of a successful car, which answered almost every criticism made about earlier models. Out went the bad habits of the old swing-axle rear suspension, and in came a much-improved swing-spring layout which killed off the old wheel camber change bogey. Out went the old gearbox, with a 'crash' bottom gear, and in came a new all-synchromesh 'box, itself a close relative of that used in the GT6. The bodyshell was completely re-skinned, tidied up in detail, given a sharply cut-off tail, and offered with a new and rather angular hardtop option; the USA-only facia of the last Mk 3s was standardized for all territories.

The 1296cc engine, though superficially unchanged, was less powerful than before, because in some ways it had been rationalized with the US-market version of the car. Although enthusiasts point out that little was changed from Mk 3 to Mk IV, independent road test figures tell their own story; both *Autocar* and *Motor* agree that the Mk IV had a lower top speed and accelerated more slowly.

In the next few years the Mk IV suffered badly in the USA market from tightening exhaust emission legislation, and was given a

above

The Mk IV announced in late 1970 was a completely re-skinned car. The rear had a much more integrated style, with a wider and flatter boot lid and with the same sort of cut-off tail as the Stags, 2000 Mk 2 saloons and TR6s of the same period. There was a completely new (optional) hardtop style.

top right

Late-model Mk IVs and all 1500s had this moulded plastic spoiler under the front bumper. It was functional if not tremendously so, and can certainly be fitted to earlier Mk IV types.

From March 1977 the Spitfire 1500 was treated to a set of cosmetic trim improvements. They were not very newsworthy – which is why British Leyland asked an ex-Miss UK, Sue Cuff, to pose with the latest car. Strangely, this is a British registered car with the North American market rubber overriders.

1493cc engine for 1973 and 1974; this was not adopted for any other market. On the other hand, there were improvements for *every* Mk IV at the beginning of 1973 – a two-inch increase in the rear track, reclining seats, smaller steering wheel, restyled instruments, a wooden-faced instrument panel, and a variety of safety-related features. The centre-lock wire wheel option was dropped during 1973, a different type of overdrive was adopted, and for 1974 a spoiler was added under the front bumper.

Mk IV gave way to 1500 in 1975, when the 1493cc engine was adopted for all markets, along with a modified gearbox having a 'single rail' selection layout. That was the year in which the same engine and gearbox were also fitted to the MG Midget 1500, something for which Triumph was never forgiven by MG fanatics, especially as the Spitfire 1500 went on to outlive the MG by a full year!

The 1500 carried on for six years, mechanically almost unchanged, though in the spring of 1977 it inherited TR7-type switchgear, and during 1978 it was given 5in rim width wheels. For its last two years it was built at Canley, almost on 'automatic pilot', and finally died in August 1980. No fewer than 314,342 Spitfires of all types were made.

Engine

Mk IV and 1500 engines were developments of those fitted to the earlier Spitfires, built in two sizes – 1296cc and 1493cc. These engines shared the same basic cylinder block, head, valve gear, pistons, and manifolding. The only important difference between the two units was that the 1493cc engine had a different, long-throw, crankshaft. All USA-market cars had a single constant-vacuum horizontal Zenith-Stromberg carburettor, while all other cars had twin horizontal SUs. Fuel injection was never used, not even in that most emissions-conscious market, California.

USA emissions regulations hit the Spitfire's power outputs very hard indeed, so much so that the 1493cc engine was introduced *for that market only* for the 1973 and 1974 model years.

The 1493cc engine was fitted to all MG Midgets built between late 1974 and 1979. Except in details such as exhaust manifolds and peripheral fixings, the two engines were virtually identical. In the 1970s, less highly tuned versions of these engines were fitted to the Toledo / 1500 / 1500TC / Dolomite 1300 / Dolomite 1500 range of family cars, which has at least ensured the continuing supply of spare parts in the 1980s.

top left

Old Spitfire 1500s sometimes lack this boot lid badge, which was a decal. Fanatical owners have been known to paint a facsimile of the decal on to the lid instead of buying a new one.

bottom left

Even late model Spitfires did not have a flat boot floor, which made it difficult (and undesirable) to carry expensive suitcases. The tonneau cover is stowed away on the left side of the boot.

Two views of a Spitfire 1500 engine bay. At this stage in the car's life, notice, the Spitfire was being produced by 'Rover-Triumph'.

As they get older, Mk IV engines of 1.3 litres behave much like those used in the earlier cars. When they were in current production the 1500s, however, were thought to be somewhat less durable, for their long-stroke design unfortunately encouraged them to knock out their main bearings, and to run their big ends. There is no permanent cure for this (except for advising the owner not to go round perpetually over-revving the unit!). However, if a 1500 unit is properly set up, properly serviced and maintained, and kept well cooled and well lubricated, this shouldn't be a problem for a 'classic' owner.

Gearbox and rear axle

For the Mk IV, there was a major improvement to the transmission. Instead of the gearbox used on all earlier Spitfires (and on the Herald range), there was an all-synchromesh gearbox, this being a modified version of that already fitted to the GT6.

Structurally, and visually, there was no difference. The gear wheels themselves and the internal ratios, however, were completely different:
GT6: 1.000, 1.25, 1.78, 2.65, reverse 3.10:1.
Spitfire Mk IV: 1.000, 1.39, 2.16, 3.50, reverse 3.99:1.

There was further rationalization for the Spitfire 1500. By this time Triumph was a part of British Leyland, technical chief Harry Webster was in charge of Austin-Rover's fortunes, and the new Morris Marina's gearbox had been a considerable re-design of that already used in the GT6.

For the Spitfire 1500, therefore (and, incidentally, for the MG Midget 1500), a version of the Marina gearbox was used, featuring that car's more compact 'single-rail' selector arrangements, but with the same internal gearbox ratios as the Spitfire Mk IV. The development wheel had come full circle.

There were, apparently, two sources of gearbox supply within British Leyland at this time, and it seems that the gearboxes fitted to the Spitfire 1500s were by no means the same as those fitted to MG Midgets or to Marinas. Logic? Don't joke – there wasn't much of that in the British Leyland empire of the 1970s.

Because the all-synchromesh gearbox was designed to accept more than the torque put out by the Spitfire engine, rather than being an old 'box modified to cope (this applied to the earlier Spitfires, as already noted), it stands up much better to sporting driving, and to old age.

When it was newly built, this gearbox design usually had quite a notchy feel; one of its characteristics was that first and reverse were quite difficult to engage when the oil was cold. This softened off as the car grew older, but by the time it was slick the synchromesh was tending to wear away. This should not have occurred before the car has recorded at least 70,000 miles (much more if the car has not habitually been driven hard), and by the same mileage the bearings and perhaps even the gear wheels and shafts might be in need of attention. The linkage could also be starting to rattle because of worn bushes and selector forks. The final 'single-rail' box was better than the original Triumph-type linkage.

If the car has an overdrive fitted, the same advice (for repairs and preventative maintenance) applies as for the earlier-model Spitfires.

As the chart concerning rear axles and their fixings has already made clear (this is detailed in the earlier Spitfire chapter) the axles fitted to Mk IVs and 1500s were much stronger than those fitted to earlier Spitfires. The 3.89:1 axle fitted to Spitfire IVs, in fact, was identical to that used to deal with the 104bhp GT6s, and the 3.63:1 axle fitted to Spitfire 1500s was no less strong. There are no horror stories to relate about late-model axles, except that they often grow noisier in old age. Reconditioned units are available, in any case, usually in exchange for the worn-out one taken off the car which you intend to restore.

Suspension, steering and brakes

Compared with earlier Spitfires, there was an important change in rear suspension to the Mk IV and the 1500 which followed it. Relatively simple changes to the central fixing (on to the rear axle housing) of the transverse leaf spring allowed it to pivot as the chassis took up a roll angle in cornering. The result of this 'swing spring' arrangement was that rear wheel camber change was drastically reduced, and the Spitfire's handling capabilities were transformed. The only other significant change, before the model ran out of production in 1980, was that the rear track and drive shafts were widened from 1973 (by one inch per side).

Except for the use of a much thicker front anti-roll bar, the front suspension, steering, and brakes of the Mk IV/1500 models were almost identical to those used on the Mk 1, 2 and 3. The steel disc wheels were of the same style throughout the 1970s, though rim widths were increased from 4.5in to 5.0in from the

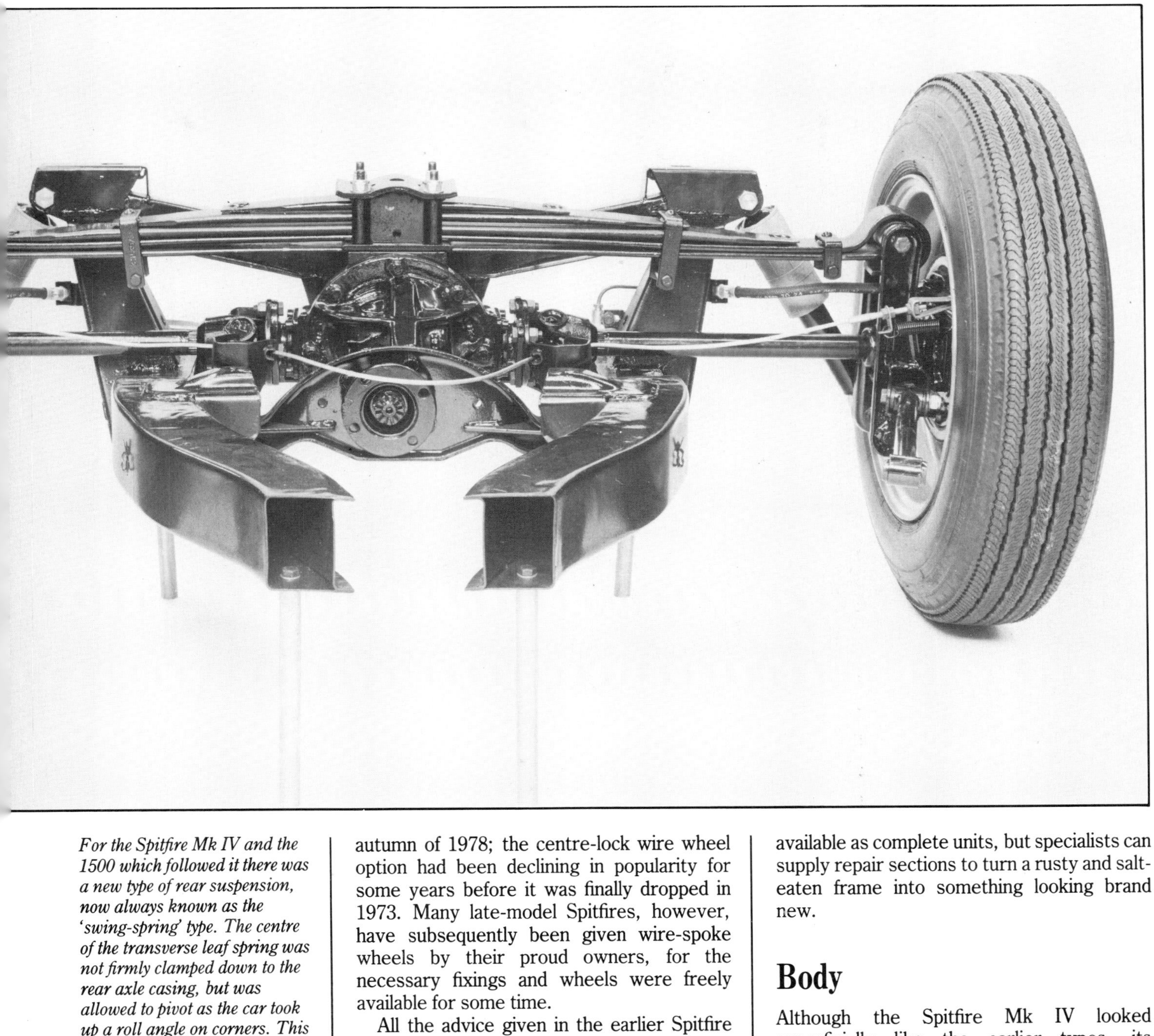

For the Spitfire Mk IV and the 1500 which followed it there was a new type of rear suspension, now always known as the 'swing-spring' type. The centre of the transverse leaf spring was not firmly clamped down to the rear axle casing, but was allowed to pivot as the car took up a roll angle on corners. This was a much better suspension than the swing-axle type, and can easily be fitted to earlier cars to give the same effect. All the parts are available.

autumn of 1978; the centre-lock wire wheel option had been declining in popularity for some years before it was finally dropped in 1973. Many late-model Spitfires, however, have subsequently been given wire-spoke wheels by their proud owners, for the necessary fixings and wheels were freely available for some time.

All the advice given in the earlier Spitfire section applies to these cars.

Chassis

The Mk IV/1500 chassis frame was almost identical to that used under the earlier types, for the 'swing spring' suspension did not involve the use of different chassis pick-ups.

The Spitfire 1/2/3 section above gives advice on checking the condition of older frames, and on repairs and restoration. New factory-specification frames are no longer available as complete units, but specialists can supply repair sections to turn a rusty and salt-eaten frame into something looking brand new.

Body

Although the Spitfire Mk IV looked superficially like the earlier types, its bodyshell had been completely 're-skinned', and there had been changes to inner panels too. In particular there was a new and smoother-profiled lift-up front end (with turned-in instead of turned-out welded flanges), different door skins with flush-mounted handles, and an almost entirely new rear end. The *basic* construction of the shell – all-steel, with a foldaway soft-top, and a detachable hardtop – was as before, but almost every detail was different. The hardtop, for instance, was a smarter but more angular style than before.

Over the years 1970 to 1980 several cosmetic facelifts took place, but these did not affect sheet metal panels.

In general, the bodies of the re-styled Spitfires (Mk IV and 1500) are similarly built, and gradually tend to rust away in the same way as the earlier type. There are advantages in owning the more modern type. Obviously these cars are significantly younger, and have therefore had less time to suffer from rain, salt and neglect and far fewer of them have spent any time off the road, deteriorating under a tarpaulin.

Although the bonnet of the later cars may *look* smoother and less prone to rotting than the earlier type, this is only because the spot-welded seam between the bonnet top and the wing pressings is internal and hidden away from view. Because of its sharp folds and seams covered by brightwork strips, the rear end of the re-styled cars was more likely to suffer from corrosion at joints, and from accidental knocks.

top left

From the mid-1970s North American market Spitfire 1500s were afflicted with these massive black rubber overriders, though British customers did not have to put up with them. The number plate holder, too, is to suit the Americans. The British must be grateful for small mercies.

bottom left

Old and scruffy Spitfires sometimes suffer badly from corrosion around the brake and clutch master cylinders, usually due to hydraulic fluid spilt on the panelwork. For restoration, a complete stripdown will be needed before re-painting can begin.

top right

This Spitfire 1500 is partly restored, but this study shows the obvious areas where filth can collect, and where water splashing can occur. On this car a new lower front body panel has had to be fitted, but has not yet been prepared for painting.

bottom right

This partly-restored Spitfire 1500 had already been treated to a lot of new body panels. One of the letters from the identifying decal has been lost. There was evidence of body rusting in most areas of the car, yet the chassis and running gear was in much better condition. The use of a separate chassis is one of the major advantages in buying a Spitfire rather than, say, a monocoque sports car such as a Midget or an MGB.

right

Worm's eye view of a partly rebuilt Spitfire 1500, whose rolling chassis was in good shape. The Spitfire body shell is very simply constructed and mainly non-stress bearing, so restoration only needs stripping out and welding expertise, new panels, and patience. The rear lower panel (below the vertical tail panel) is in the line of fire for everything thrown up by the rear wheels.

right

Most Spitfires suffer corrosion on the rear wings ahead of the line of the rear wheels. This car has already had a repair panel welded into place, and the sills have already been renovated.

left

Body restoration on this Spitfire is well advanced at Chiswick Car Craft in West London. New outer sills have already been fitted, and because of the offside rear wing was completely rotten a new panel is about to be welded into place. The inner wing was in surprisingly good condition, though work was needed around the door closing panel itself.

Not too much rust inside the boot area of this Spitfire, which has clearly not suffered too much from water leaks around the boot lid seals. This part of the body shell is scruffy but basically sound, and will present no restoration problems.

Another view of the jigsaw of metal pressings around the rear corner of a Spitfire Mk IV or 1500. When the rear wing is welded into place the whole area will be ready for finishing off and painting.

On this Spitfire it was necessary to make repairs to the floor panel under the seats and close to the B/C post, but the panelling adjacent to the rear suspension radius arm pick-up points was still in excellent condition. On some cars this floor panel rusts through – replacements, fortunately, are available.

On Spitfires, rusting often occurs near the sharp corners surrounding the boot lid. This is encouraged by water collecting in the gaps between the rubber seals and the panels themselves. Welding work has already been carried out on this car, to make good before other panels are welded on again.

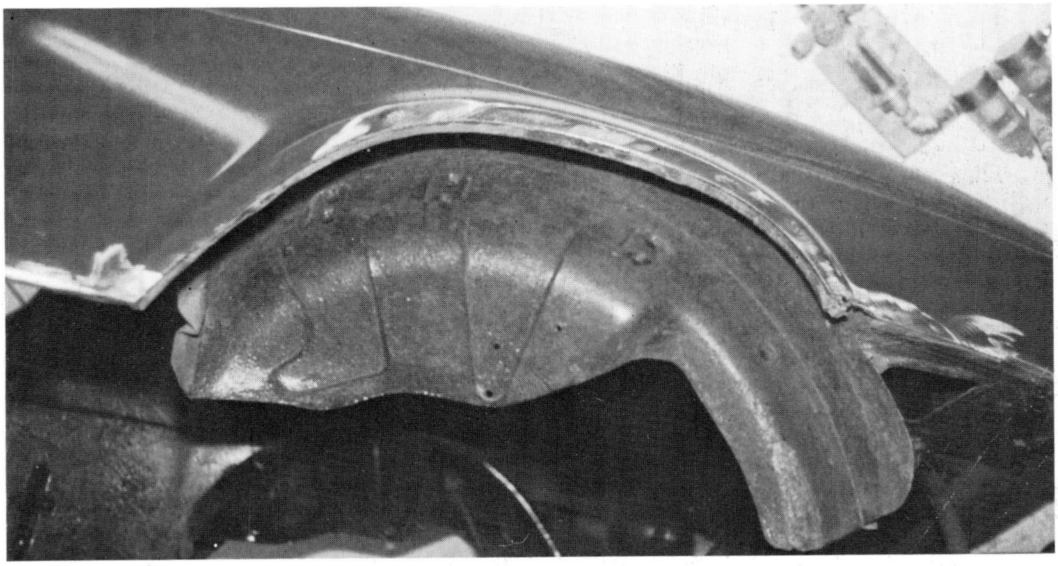

On any Spitfire or GT6, always look closely for corrosion around the edge of the wheel arch cutouts. Repair work has started on this car, which looked very 'frilly' indeed before restoration work began.

Such major sill and lower scuttle repair work is a very common job on Spitfires. The rear wing has already had to be replaced – which is another common task. This car was eventually returned to near concours condition.

Look at the awful state of the front lower corner panel of this 1972 car – actually a GT6, but the construction is just the same on Spitfires. A replacement part, held out for inspection by our charming model, will be needed. A lot of reclamation work is also needed in the headlamp nacelles. In both cases, standing water and general old age are the culprit.

The GT6 concept started life in 1963 as a Spitfire GT proposal from Michelotti, but was turned into a six-cylinder hatchback coupé before being launched in 1966.

From this view it is easy to see why some observers dubbed the GT6 a 'mini E-type'. It sounded the part too!

Except for the prominent bonnet bulge, there was no obvious clue that Standard-Triumph had shoehorned a six-cylinder engine into an engine bay originally designed to take a four. As on the Spitfire, wire wheels were an optional extra.

Although the GT6's driving compartment was the same size as that of the Spitfire, it had been completely re-designed, with a fully-equipped polished wood facia, and more plush carpets and seat designs. The steering wheel on this car is non-standard.

Triumph squeezed the long, slim, six-cylinder into place by using a compact cooling radiator mounted well forward into the nose. The flame-trap air-cleaners on this engine are non-standard.

Replacement of the outer sills is well under way. Because the sills provide much beam strength in a Spitfire body, it is essential that the door opening should be supported while this is going on. And what better support is there than the door itself?

Interior

In its appointments and fittings the Mk IV (and the 1500 which followed it) was a much more up-market car than the Mk 3 which it replaced. The USA facia style had been adopted for all markets, there was an alloy flat-spoked steering wheel, a steering column lock, larger and more plush looking seats, and better quality carpets and trim panels. The boot was larger than before, and (being squared up) was a more convenient shape.

Cosmetic and equipment facelifts were introduced in 1973 (reclining front seats being the most important item), and again in 1977 (TR7 steering column switchgear, and yet more changes to the seats); when the 1500 appeared at the end of 1974 it was also treated to a polished wood facia panel.

Restoration

It is easier to complete the restoration of a Mk IV or 1500, simply because the parts are more easily available, and because a larger proportion of these cars are still on the roads. Body panels, in particular, are easier to come by, and much of the re-manufacturing effort has been concentrated on this, rather than the earlier, type.

All the general advice given regarding restoration (and the sequence of work you should follow) has been listed in the Spitfire 1/2/3 section.

Maintenance

For all relevant advice, see the earlier Spitfire 1/2/3 section.

Cost guidelines

Except that they received re-skinned/facelifted body styles, all-synchromesh gearboxes, and slightly modified rear suspensions, these types of Spitfire were basically the same as the earlier-type (Mk 1, 2 and 3) models already surveyed. There are, however, two basic reasons why the re-skinned cars are more desirable and command higher prices today. The obvious one is that they are, by definition, newer cars (which means that most will not have deteriorated as badly, or need as much restoration), but the other is that they are better cars in every way.

A good Mk IV commands much the same sort of price as a good Mk 3, but the final Spitfire type, the 1500, is significantly more valuable than any other type. Expect to have to pay at least one-third more for a 1500, and for the price to be up to twice as high as that of a Mk 1. A good 1500, by the way, is almost as valuable as a good GT6 Mk III.

As with the earlier types, wire wheels, a hardtop and overdrive are all desirable extras. Don't forget that wire wheels were officially dropped from the options lists in 1973, though some cars have had them added retrospectively.

In a nutshell

Without a doubt the Mk IV and 1500 are the best and most desirable Spitfires of all. Most people agree that the re-styling exercise, which ushered in the Mk IV in 1970, was a real success. It combined the original charm of the early-type style, with more modern elegance, a family likeness with other sporting Triumphs of the day, and a better trimmed and furnished interior.

Compared with Spitfires 1, 2 and 3, the Mk IV had nicer lines, a more capacious luggage boot, a smarter (optional) hardtop, a modern transmission and vastly improved rear suspension. The 1500 which followed it at the end of 1974 was an altogether faster, lustier, and more lively car with higher gearing; unfortunately for Triumph enthusiasts, this was the Spitfire which should have been launched years earlier, and the pity of it all was that no other mechanical improvements were made to take the car into the late 1970s. On the other hand, make no mistake, the definitive Spitfire was a much better car than the MG Midget of the 1970s, something which MG enthusiasts could not forgive British Leyland's bosses, or admit in public!

This facia style was first used on late-model Spitfire Mk 3s sold in the USA, but was fitted to all Mk IVs. For these cars it had a vinyl-covered panel. Compare it, however, with . . .

. . . the wood veneer panel of the Spitfire 1500, which came along in 1974. Note in each case the location of the overdrive switch which was in the gear lever knob. Previous Spitfires had overdrive control stalks on the steering column.

SPECIFICATIONS	Spitfire Mk IV	Spitfire 1500
Engine	In-line four	In-line four
Construction	Cast-iron block, cast-iron head	Cast-iron block, cast-iron head
Main bearings	Three	Three
Bore/stroke	73.7 x 76mm	73.7 x 87.5mm
Capacity	1296cc	1493cc
Valves	Pushrod ohv	Pushrod ohv
Compression ratio	9:1	9:1
Fuel system	Two SU Type HS2 carburettors	Two SU Type HS4 carburettors
Power	63bhp (DIN) at 6000rpm	71bhp (DIN) at 5500rpm
Torque	69lb ft at 3500rpm	82lb ft at 3000rpm
Transmission	Four-speed manual, optional overdrive on top and third	Four-speed manual, optional overdrive on top and third
Top gear	16.7mph at 1000rpm in top, 20.8mph at 1000mph in overdrive top	18.0mph at 1000rpm in top, 22.6mph at 1000rpm in overdrive top
Final drive	Hypoid, 3.89:1 ratio	Hypoid, 3.63:1 ratio
Brakes	9in Girling front discs and 7 x 1.25in rear drums	9in Girling front discs and 7 x 1.25in rear drums
Suspension, front	Ind. by coil springs, wishbones and anti-roll bar, telescopic dampers	Ind. by coil springs, wishbones and anti-roll bar, telescopic dampers
Suspension, rear	Ind. by 'swing spring' transverse leaf spring, swing axles, telescopic dampers and radius arms	Ind. by 'swing spring' transverse leaf spring, swing axles, telescopic dampers and radius arms
Steering	Rack and opinion	Rack and pinion
Wheels/tyres	Steel disc wheels, 4.5in rims, or optional centre-lock wire spoke wheels, 4.5in rims. 145-13in tyres	Steel disc wheels, 4.5in rims, or optional centre-lock wire spoke wheels, 4.5in rims. 155-13in tyres
Chassis	Pressed steel backbone style chassis frame, with box section side members	Pressed steel backbone style chassis frame, with box section side members
Body	Pressed steel bodyshell, with two seats and two passenger doors. Open sports car, with optional detachable steel hardtop. Entire nose section lifted up for engine bay/suspension access	Pressed steel bodyshell, with two seats and two passenger doors. Open sports car, with optional detachable steel hardtop. Entire nose section lifted up for engine bay/suspension access

DIMENSIONS		
Length	12ft 5in	12ft 5in
Width	4ft 10.5in	4ft 10.5in
Height	3ft 11.5in	3ft 11.5in
Wheelbase	6ft 11in	6ft 11in
Unladen weight	1717lb	1750lb

GT6 Mk 1, 2 and 3

The Vitesse 2-litre of 1966 used exactly the same engine and gearbox as the original GT6, and the suspensions were closely similar too.

With the original Spitfire and the six-cylinder Vitesse both being put on sale in 1962, it didn't take long for the pundits to decide that a six-cylinder Spitfire would soon appear. But it didn't happen in a hurry – the original GT6 was launched four years later.

In its basic layout the GT6 used the Spitfire's chassis layout married to a 1998cc six-cylinder engine, and Spitfire's bodyshell modified to become a fastback/hatchback coupé. There were, of course, many differences in detail.

Let's put it this way – the Vitesse 2-litre evolved from the Vitesse 1600, which had itself evolved from the Herald. The GT6 used the same engine, transmission and brakes as the Vitesse 2-litre. The Spitfire's chassis and suspension had evolved from that of the Herald. Got it?

There were problems, of course. Because the six-cylinder engine was heavier and relatively further forward, the GT6 had a different weight distribution from the Spitfire (more of the weight was over the front

The GT6 was Triumph's first-ever hatchback. Restorers should notice the location of the rear badge, and the fuel filler cap, which is actually the same as that used on Spitfires but in a different position. Wire spoke wheels were optional.

wheels). Not only that, but the car was initially aimed at a rather less sporting market, which meant that it was given rather softer springs and damper settings.

The original GT6 (later dubbed Mk 1) had a 95bhp engine, and a newly-developed all-synchromesh gearbox, allied to a much stronger axle than that used on the Heralds and Vitesses. Compared with the current Spitfires (Mk 2, and, shortly, the Mk 3) it had more plush seats, an alloy-spoked steering wheel, and a wooden facia panel, plus opening rear quarter windows to help the car's ventilation. Most cars had overdrive, but this was (and always remained) an option.

Two years after the GT6 had appeared, it was re-launched as the much-improved Mk 2 model; in the USA it was badged as 'GT6 Plus'. All mechanical improvements were also shared with the Vitesse 2-litre Mk 2. To cure the first car's wayward handling, there was a new rear suspension, featuring a reversed lower wishbone and a flexible 'rubber doughnut' in the drive shaft.

At the same time, the TR5-type cylinder-head casting was adopted, and peak power rose to 104bhp; in the next five years there were few engine changes, except to keep ahead of exhaust emission rules – in the UK this trimmed peak power to 95bhp by 1972, and in the USA to a mere 79bhp by 1972. This was all very disappointing for Triumph's engineers, who saw the performance of their sports cars being whittled away quite out of their control. There was a new facia style with matt instead of gloss wood finish, much

top right

The GT6 Mk 2 proclaimed its presence with different badging, now fitted to the right of the rear panel. This car, well spattered with mud after a test drive, shows where filth can collect behind the rear wheels and cause corrosion to start.

bottom right

A well-known publicity shot of the 1969-model GT6 Mk 2, taken on the sports ground at the Triumph factory, showing the styling changes made from Mk 1 to Mk 2 – the mock-Rostyle wheel covers, the extra cooling louvres in the side of the bonnet, and the extra air outlets for the cabin behind the rear quarter window.

left

GT6 Mk 1 on the standard disc wheels.

below

When the GT6 became Mk 2, in the autumn 1968, it was given the same 'bone-in-teeth' bumper as the Spitfire Mk 3 and some rather pseudo-Rostyle wheel trims.

improved cabin ventilation (with exit louvres behind the quarter windows), the 'bone-in-teeth' front bumper layout of the Spitfire Mk 3, and different wheel trims.

All in all the Mk 2 was a much better car, now with a top speed of nearly 110mph, with acceleration and handling to suit.

Finally, at the end of 1970, the Mk 3 was announced. The arrival of this type was inevitable, as it featured all the changes to body construction and styling which were a part of the Spitfire Mk IV. To match this, the profile of the rear quarter windows was changed. As with the Spitfire Mk IV, the screen was two inches deeper. The overdrive control (where fitted) now found itself in the gearlever knob, and there was a different type of steering wheel – each GT6 derivative, in fact, had its own style.

At the beginning of 1973 the GT6 received one last revision, when the 'lower-wishbone' type of rear suspension was dropped in favour of the latest wide-track swing-spring Spitfire Mk IV type; just 4218 of the swing-spring GT6s were produced before the model was dropped at the end of 1973. A total of 40,926 cars of three distinct types had been produced in seven years.

TRIUMPH
GB
MWK 670G

above

Inboard of the rear wing chrome strips and below the top edge of the rear panel the GT6 Mk 3 and the Spitfire Mk IV models used the same body pressings.

left, both pictures

The re-skinned GT6 Mk 3, with neatly cut-off tail and more stylish weels (these were also fitted to the last Mk 2s as well). The bonnet panel now has welded seams along the flanks rather than on top of the wings, there are differently shaped louvres behind the rear quarter windows, and those quarter windows have a different shape

Engine

The GT6's engine was a close relation of the four-cylinder Spitfire unit, with the same basic design, some common components, and a modicum of shared machining facilities in Coventry.

The original straight-six was shown in 1960, for use in the Standard Vanguard Six saloon. It was, in effect, a six-cylinder version of the four-cylinder 'SC' unit, and had four crankshaft main bearings. With a bore and stroke of 74.7 × 76mm, its capacity was 1998cc, this being the size used in all GT6s built from 1966 to 1973.

Although this engine was not as widely used as the little four-cylinder engine it still found a home in many different models. In smaller (1596cc) guise it was used in the original Vitesses of 1962-1966, and in enlarged (2498cc) form it was used in cars like Triumph TR5/TR6 sports cars. 1998cc and 2498cc engines figured in the line-up of handsome Michelotti-styled saloons such as the 2000 and the 2.5PI, built from 1963 to 1977.

Like the Spitfire engines, all GT6 units had simple overhead valve gear, with a block-mounted camshaft driven by chain from the nose of the crankshaft. Cylinder blocks and heads were made in cast iron; two completely different cylinder head castings were used – one for the GT6 Mk 1, the other for succeeding models.

The original GT6 unit was nearly identical to that used in the contemporary Vitesse 2-litre but more different from that used in the

Triumph 2000 models. The cylinder head design was like that of other Standard-Triumph straight sixes, which is to say that the porting was not as efficient as might have been wished. Even so, this engine produced 95bhp, which was enough to give the GT6 a 105mph top speed.

When the engine was first installed in a TR sports car – the TR5/TR250 range of 1967 – it was not only enlarged to 2498cc by the use of a long-stroke crankshaft, but given a new and much deeper-breathing cylinder head. This head was donated to the GT6 Mk 2 of 1968 (along with the commonized cylinder block which coped with 1998cc and 2498cc capacities), which allowed power to be pushed up to 104bhp, and the car's top speed to 110mph. This head, incidentally, needed new inlet and exhaust manifolds to suit.

Thus equipped, the GT6 carried on to the end of its career, with only minor changes from year to year (and, of course, to suit USA-market exhaust emission rules). Many components in this engine were shared with other cars using one or other version of the straight-six unit, as a comparative study of the Parts Catalogues will confirm. However, it is worth emphasizing that although all the 'full-width' cylinder-head *castings* were the same, the machined cylinder heads were by no means identical, for different size combustion chambers were used on different models. Components such as the timing gear and details of the valve gear were, however, common to all the six-cylinder engines.

Like the Spitfires, these GT6s have simple engines which should never be beyond the understanding of a competent mechanic. Some of the old-age problems are shared with the Spitfire – notably the rather busy-sounding valve gear, even when the engine is in good condition, noise which gets progressively more insistent as wear sets in.

By about 70,000-80,000 miles a GT6 engine needs attention to its timing chain (which may be stretched, or its tension out of adjustment), and perhaps to its water pump. If the engine is being stripped out for such work to be done then you should also give it a head-off top-end overhaul to reduce oil consumption. Healthy examples should have 40 to 60 psi oil pressure in normal running. Twin SUs may need re-balancing, but this is simple enough if you follow the correct procedures.

above

The original GT6 engine of 1966 still used a dynamo to generate electricity. Note how the engine's electrical components are all grouped on the near side of the unit, with the carburettors on the other side.

top right

By 1971 the GT6 Mk 3 was equipped with a different cylinder head casting (note that there are now no separate pushrod cover tubes between the sparking plugs) and a large alternator.

bottom right

The engine bay of all GT6s (this is a Mk 3) was well-filled. In this view the carburettor air cleaner looks vulnerable to attack from water splashed up from the front wheel. However, when the bonnet assembly was closed the inner wheel arch pressing covered it.

Some GT6s tended to overheat, particularly the Mk 1 examples where the water-coolant flow in the head was not ideal, and the radiator capacity was somewhat limited. Adding an accessory electric fan (there is *just* space for a slim-line Kenlow-type fitting) often works wonders. It is well known that the early engines had a very high octane requirement; since such exotic fuels are no longer available you should keep a very careful eye (and ear) on its behaviour, especially for pre-ignition.

Gearbox and rear axle

For the GT6/Vitesse 2-litre models, Harry Webster's team developed a new four-speed all-synchromesh gearbox. This, in prototype form, had been used by the 'works' racing and rally cars in 1964-1966. It looked very similar to the normal Herald/Spitfire 'box (and used the same remote control mechanism) but had a longer cast-iron casing, and mainly different internals. Top, third and second gear ratios (but not all the components) were the same as on the Vitesse 1600, but first and reverse gears were different. D-type overdrive, as used in the Spitfires and earlier Vitesses, was still optional.

This was the gearbox which also formed the basis of the all-synchromesh unit later fitted to Spitfire Mk IVs and 1500s, and it was also used in cars like the Dolomite 1500 and 1850 – in most cases with different internal ratios; the Dolomite Sprint, on the other hand, used the larger TR6/Stag type gearbox. One major problem had been solved. Unlike the earlier Spitfires, there was no problem with jumping out of mesh in reverse gear.

Except that the new gearboxes had a rather notchy feel (Triumph 'boxes – notably those used in the TR4/TR5/TR6/Stag family – all seemed to feel the same in the 1960s and early 1970s), these were good, solid, units. First and reverse gear were often difficult to engage when the transmission oil was cold and thick – this, and the general notchiness, tended to improve as the gearboxes grew old and began to wear.

Don't expect to do much work on this gearbox until the mileage climbs past 60,000-70,000, at which point you might need to have new synchromesh cones, and perhaps bearings (and even gear wheels) changed. A sportingly-driven GT6 gearbox will also exhibit a worn and rather rattly linkage, with the change less precise than you might hope.

Overdrives (fitted to a high proportion of all GT6s) work well as long as their lubricant is clean and their external electrics are in good condition. Naturally they have to work harder on this car than on, say, the Spitfire Mk IVs and 1500s, but they stand up well to their job.

As with the late-model Spitfires, reconditioned exchange gearboxes and overdrives are freely available.

The axle fitted to the GT6 and, of course, to the Vitesse 2-litre was purpose-designed for the job of transmitting 100bhp, and is well up to its task. Compared with the earlier (Herald/Vitesse 1600) type of axle, it is considerably stronger, and should give little trouble in an older, classic GT6. It may, of course, have grown somewhat noisy but, as with the rest of the transmission, it can be replaced by a newly reconditioned unit.

The chart of information included in the Spitfire 1/2/3 section shows why the GT6 axle is so much more robust than earlier types. It also shows that it has much in common with the axles fitted to Spitfire IV and 1500 models.

Then there is the question of ratios. Most cars had 3.27:1 without overdrive, or 3.89:1 with overdrive. *Some* Mk 2s had a 3.27:1 ratio even if overdrive was fitted, but in my opinion this made the car much too high geared; when the time comes for a rebuild – and you want to keep the car original – I reckon that you should revert to a 3.89:1 ratio (but don't forget to buy a replacement speedometer at the same time . . .). The Spitfire 1500's 3.63:1 ratio would fit, of course, but that would make the car non-standard, and in any case there is no shortage of 3.89:1 ratios at the moment.

Suspension, steering and brakes

The GT6's front suspension and steering were similar to those used in all the other Herald/Vitesse/Spitfire family of cars. Because the GT6's engine was much heavier than that of the Spitfires, and because there was a higher proportion of the weight over the front wheels, the spring rates and damper settings were unique to this car. As on the Spitfires, the rack and pinion steering was inch-accurate, but very low geared – 4.25 turns lock-to-lock.

All the advice given for the checking and rebuilding of Spitfire front suspensions applies equally to the GT6. In fact the GT6 gives its front suspension a harder time, mainly because the static load is about 550lb per wheel, compared with 480lb per wheel on a Spitfire of the same age. This means that wear in bushes, and especially in vertical link trunnions, will occur even more persistently than in the Spitfires.

A GT6 Mk 3 at 60mph in overdrive top, with the engine turning over at only 2500rpm.

On a neglected GT6, as with Spitfires of a similar type, the trunnions will almost certainly be badly worn. Fit new trunnions, then swear never to neglect their lubrication in future. Consult the Workshop Manual for details of the grade to be used.

Compared with the Spitfire, GT6 springs and dampers had somewhat softer settings, but don't be fooled by a car which feels distinctly 'floaty' when you try it out. Dampers, in particular, may be worn. Unless you already are a GT6 expert, it is advisable to find such a paragon and take him along with you when you want to assess a car.

On the GT6 Mk 1 the rear suspension was of simple swing-axle type structurally identical to that used on early type (Mk 1, 2 and 3) Spitfires; once again, the spring rates and damper settings were unique to this model.

Because there was strong criticism of the original rear suspension, it was completely re-jigged for the GT6 Mk 2, and the same set-up was used on Mk 3s up to the beginning of the 1973 model year. The basic chassis was not changed, nor was the position of the transverse leaf spring.

However, a 'reversed' lower wishbone linked the frame side member to a new

vertical link and there was a rubber 'doughnut' in the drive shafts to accommodate plunge. The position of the radius arm forward pivot point was moved to help provide better geometry.

The overall result was that there was a drastic reduction in wheel camber change, and the car handled much better than before. The same basic suspension was used on Vitesse 2-litre Mk 2 models, except that these cars used lever-arm dampers instead of telescopics.

There was one final change. Probably in the interests of cost-saving the 'lower wishbone' scheme was abandoned for 1973, to be replaced by the wide-track version of the Spitfire IV/1500's swing-spring system instead.

GT6s suffer from the same 'growing-old' problems as the Spitfires, only more so, and rather earlier. This is because the cars are heavier, and because there is much more torque being handled by the transmission. In the case of Mk 1 and late-model (1973-specification) Mk 3s, the usual 'consumables' such as bushes at spring, damper, and radius arm mounting ends will have to be inspected regularly. In addition, universal joints in the drive shafts (and in the propeller shaft, too) wear well before middle age is reached, become noisy, and should be replaced.

GT6s fitted with the 'reversed lower wishbone' type of rear suspension handle a lot better than the Mk 1s ever did, but bring their own problems with them. In particular, you

Put a Mk 2 or pre-1973 Mk 3 GT6 on a ramp, look up at the rear suspension, and this is what you will see. Instead of the simple swing-axle suspension of earlier cars, there is a reversed lower wishbone (a forged item) with a single pivot on the chassis frame, and two on the vertical link holding the wheel hub. There is a large Rotoflex coupling in the driveshaft to accommodate torque and *plunge, and the telescopic damper is behind the line of the driveshaft.*

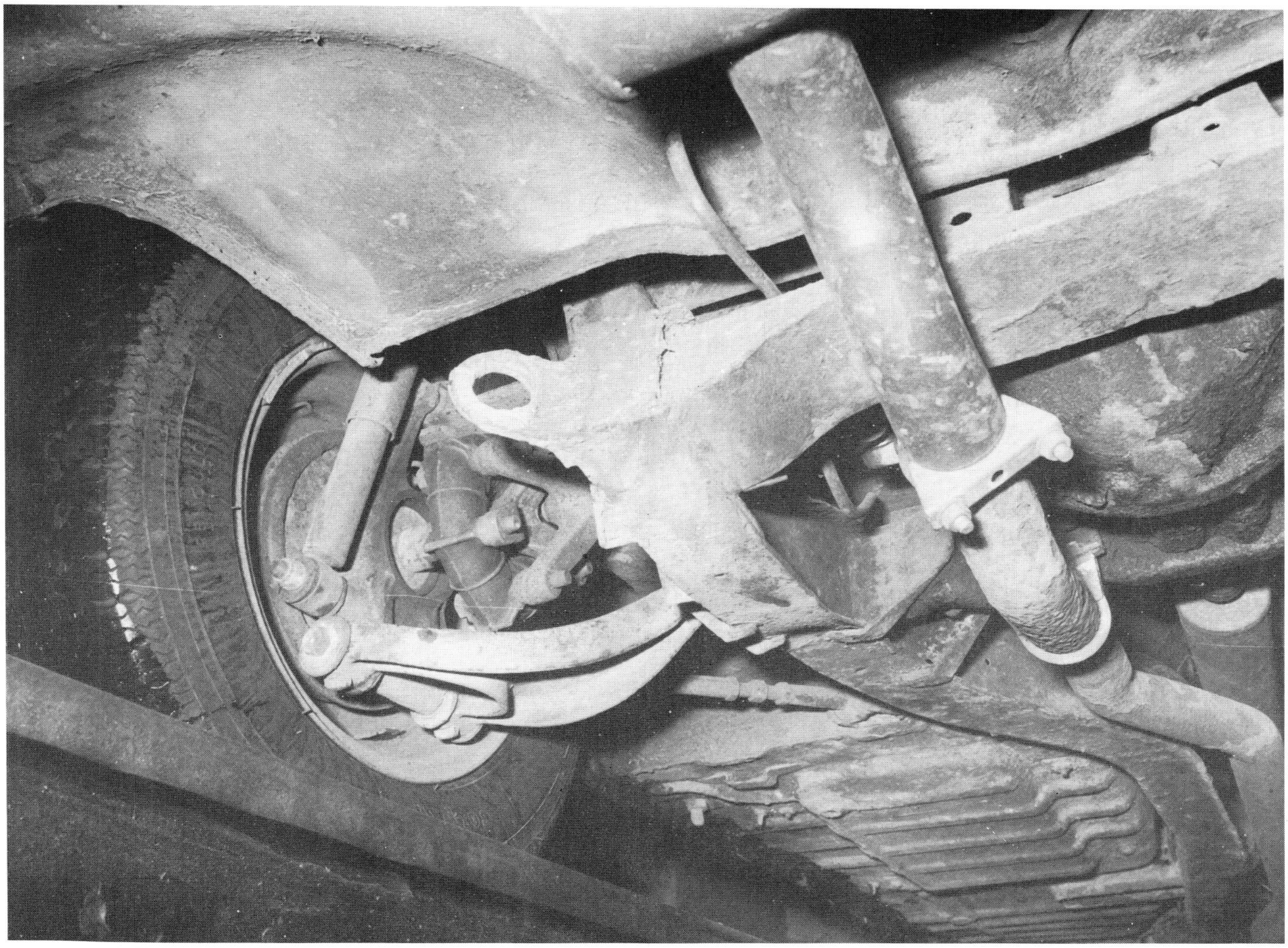

On all Spitfires and GT6s, as exemplified by this shot of a GT6 Mk 2, there were dozens of nooks and crannies where water and road salt could collect. A really good under-car inspection is needed before you purchase any car.

will need to check the condition of the Rotoflex coupling in the drive shafts, which not only has to provide torsional flexibility, but also has to accommodate 'plunge' in the shaft length as the suspension moves up and down. The coupling itself tends to crack up around the bolt holes, and the rubber may begin to peel off its bonding with the metal; damage from stones or other debris thrown up by the wheels can also affect its integrity.

The bushes in the reversed lower wishbone also wear, become noisy, and destroy the precise geometry of the suspension, but at least they are readily available, and simple enough to replace.

Like the contemporary Spitfires, the GT6 was offered with steel disc or centre-lock wire spoke wheels. In each case these had 4.5in rims, and — like the Spitfire — the disc wheel type changed in 1970 to a rather attractive exposed-nut variety. The wire-spoke wheels have a harder time than they do in the Spitfire — a combination of extra weight and extra power to be accommodated — so it is doubly important to check up, from time to time, for loose or bent spokes, and for worn centre-lock splines.

The GT6's brakes were larger than those fitted to the Spitfire, and some later models had a vacuum servo as standard. Provided the car has not been allowed to stand unused for a long time (in which case the discs will have become rusty), there should be few problems on older cars. As with the Spitfires, though, the handbrake cables tend to rust up and seize.

Chassis

As already made clear, the GT6's chassis frame was almost identical to that of the Spitfire, except for the location of brackets at the front end suitable for mounting the longer six-cylinder engine and the relocated water radiator.

When the GT6's rear suspension was changed from swing-axle to the 'lower wishbone' type, the only important chassis change was to provide a pair of substantial new mounting brackets for the inner pivots of the reversed wishbones. With this suspension, in fact, the chassis mounting for the rear dampers became redundant, as the dampers were fixed to the inside of the rear wheel arch.

A GT6 frame eventually suffers from the same problems as that of a Spitfire.

Body

In effect, the GT6 had a very cleverly re-engineered version of the Spitfire bodyshell, with all the important changes concentrated at the rear and over the driver's head.

At the front of the original car, the main differences were the provision of a power bulge in the top of the bonnet (to clear the six-cylinder engine, which was mounted relatively further forward in the chassis than the smaller four of the Spitfire), and the use of a different radiator grille. The doors were modified to include front window frames and swivelling quarter windows, but used the same basic lower pressings.

The unique feature of the GT6, of course, was its long sloping coupé roof, which incorporated a lift-up hatchback (or, as the Americans tended to say, 'glass-back'). This was incorporated without changing the windscreen, outer rear wing, or rear panel pressings. Opening rear quarter windows were fitted immediately behind the line of the doors.

Structurally, however, the rear end of the body had to be changed. On the Spitfire the petrol tank was mounted above the line of the rear suspension, behind the seats, but on the GT6 a new tank was located on the left side of the boot floor, with the spare wheel moved over to the right side to accommodate it. Instead of being behind the seats, on the rear deck, the fuel filler was located on the tail, just below the hatchback aperture.

The original, rather rounded, tail was used in GT6 Mk 1 and 2, the raised front bumper (Spitfire Mk 3 type) being a Mk 2 recognition point. When the Spitfire became Mk IV, with a complete re-skin, the GT6 followed suit. Accordingly, the Mk 3 had the somewhat sleeker front end with hidden panel flanges, the recessed door handles and the cut-off tail of that car. At the same time the shape of the rear quarter windows was 'softened', and the fuel filler cap position was moved to the left side rear wing panel. In Mk 3 form this was an extremely elegant and carefully detailed style which is appreciated more today than it was at the time.

Rust and deterioration in a GT6 occur in the same places as in Spitfires of the period. In addition, there may be rusting on the roof

This GT6 Mk 2 looked like a 'basket case' before restoration began, but Chiswick Car Craft eventually brought it back to life. Most skin panels needed to be renewed and the outer sills, door skins and rear wings have already been tackled. Note the grit blasting on the bulkhead behind the front wheel.

Lift the hatchback of any GT6 (this is a Mk 3 model), and you will see a very spacious luggage platform – but there is no attempt at providing useless occasional seats.

The coupé top of the Mk 3 GT6's new bodyshell had re-profiled rear quarter windows and modified cabin air outlets. The fuel filler cap was neatly faired into the left-side rear wing.

New noise regulations meant that the Mk 3 GT6 needed a large transverse silencer in its exhaust system, visible from this but few other angles. Mk 3s were produced from late 1970 to the end of 1973, and would achieve 110mph.

Except for the much larger and more expensive Stag, the GT6 Mk 3 probably had the most thoughtfully-equipped facia and instrument panel of any early 1970s Triumph. Note the 'eyeball' face-level ventilation outlets.

Late model GT6 engines, like 1970s-style Spitfires, drew their intake air from the front of the car through flexible tubes. Compared with similar Spitfire engine-bay shots, the extra length of the six-cylinder engine is obvious.

Parts of this GT6 Mk2 could be renovated (such as the roof panels and the hatchback), but the complete door has already been replaced, and a new rear wing will soon be welded into place. New inner and outer sills have already been fitted. To keep the body as stiff as possible, tackle panel replacements one at a time – in this case it was not really desirable for the door closing plate (B/C post) to be removed at the same time as the rear wing, but there was so much corrosion that this was considered necessary.

A new front wing panel (which is part of the bonnet assembly, of course) was needed, but the rest of the bonnet and the front lower panels could actually be reclaimed on this car. To an accomplished sheet metal worker, 'unstitching' a panel from its neighbours is quite a straightforward business.

Making do and mending: when this GT6 Mk 3 was rebuilt the proper left rear wing was not available. A Spitfire wing panel was fitted, then re-fettled to accept the fuel filler neck which is a feature of this type of GT6. The rest of the body shell has been restored rather than panels being replaced.

left

More make do and mend. Above the welding line the panels have been taken back to bare metal and the corrosion removed, preparatory to a full repaint. Below the line, however, a new rear lower panel has just been fitted.

below

Battery acid, spilt down the sides of the battery had caused severe corrosion to the pressed support panel, so in this restoration it was necessary to 'stitch' a new sub-assembly into place. A lot of tidying up and sealing remains to be done before this section of the car can be repainted.

panel, close to the guttering at the front (above the screen) and sides (above the doors) – this is usually worse if a GT6 has spent much of its time standing outside, uncovered. Rust may also occur on the transverse panel seam across the tail (Mk 1 and Mk 2 cars), and around the brightwork edges to the cut-off tail of the Mk 3. If the tailgate leaks water, this may affect the metal around the edges of the hatchback, and as the water collected under the fuel tank, the floor panel under the tank.

Interior

As with the Spitfires, there were several different generations of trim and facia furnishing in the GT6s. The original cars had a bright polished wood facia panel, and an aluminium-spoked steering wheel, slightly dished. Mk 2 cars had a different instrument layout, a satinized wood facia finish, and re-located heater controls, plus eyeball-type face-level fresh-air vents, along with the padded steering wheel spokes. Reclining seats and a flat-spoke steering wheel were standardized on Mk 2s from mid-run, in the autumn of 1969.

Mk 3s had a similar facia layout, the flat aluminium-spoked steering wheel was continued, and the seats had longitudinal rather than transverse pleating. The overdrive control, which had been on the steering column in Mk 1 and 2, found itself in the gearlever knob on the Mk 3.

Because the GT6 had such a compact little cabin – well, we might as well admit, it was a rather cramped cabin – it was easy for the trim panels to be scuffed or damaged. The fact that the hatchback door was so large also encouraged owners to thrust outsize loads into the back. As with the later-model Spitfires, therefore, it's nice to know that replica trim panels are now available, along with carpets and rear floor coverings; until very recently the replica carpets were not moulded in the fashion of the original factory items, so they might not fit quite as well as you would like. Changing the front carpets is only possible after you remove the scuttle brace, and since that might also mean disconnecting the radio fittings you might be reluctant to tackle this at first. Do persevere – the result is well worth it.

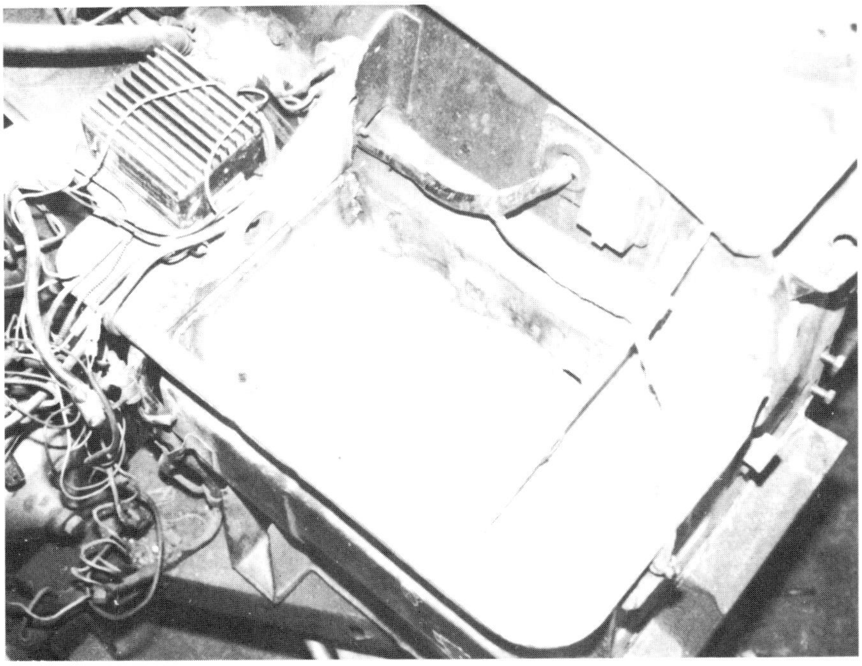

top right

The facia of the original GT6 had a gloss-finished wooden panel and a full display of instruments, plus an alloy-spoked wheel. The front doors had wind-up windows and opening quarter lights.

bottom right

There was a little space for stowage of valuables in the GT6 behind the seats and under the load platform. Not lockable, though.

Look carefully at the condition of the wooden facia panels. The veneer is thin, and it is all too easy for the panels to be scratched. There is really no way for scratches to be repaired, and I would not recommend trying to disguise them with any form of staining. With this, and with any discoloration which has taken place over the years (strong sunlight and regular hot weather eventually have an effect), it is best to take a deep breath and acquire a replacement.

Restoration

Although there are fewer GT6s than Spitfires on British roads, much of the restoration expertise is shared with the Spitfire. Any specialist company or workshop that can deal with a Spitfire can certainly tackle a GT6 as well. Mechanical parts are freely available, as are many of the body panels.

All the general advice given regarding restoration priorities for Spitfires applies equally to the GT6s.

This well-used GT6 Mk 2 test car shows how necessary were the heel pads under the driver's feet. These eventually wear through, and may make a car look extremely tatty in old age. Like all other trim items on these car replacement items are available. Note the dirt deposits on the inside door pressings on the far side of the car – corrosion eventually sets in there.

Maintenance

In general, a GT6 should be maintained according to the same standards as the Spitfire, perhaps with extra attention being given to the powerful six-cylinder engine and its rather complex cooling system. (For all relevant advice, see the earlier Spitfire 1/2/3 section.)

Cost guidelines

As you might expect from all the information given in this book, the GT6 is not to everyone's taste. The Mk 1's reputation suffered at the hands of the motoring media, and has never truly recovered from that onslaught. In any case, *all* GT6s had to compete for attention with the MGB GT, especially in the United States market – although the GT6 was as fast and as well-equipped as its rival from BMC it had much less space in the cabin.

These days, although a GT6 Mk 1 or Mk 2 sells for very little more than a Spitfire of the same period, I think it is intrinisically worth considerably more than this. However, the price commanded by a Mk 3 is always more – perhaps 50 per cent more – than that of the equivalent Spitfire (the Mk IV), and that seems to be a much more equitable statement of value.

Like the Spitfires, wire-spoke wheels and overdrive add a little – £100/£200 – to the value of a GT6.

Jack and tool kit of the GT6. The spare wheel, jack and tools were all normally covered by a trimmed board, which has been removed for the picture to be taken.

This wide-angle camera shot rather flatters the packaging of the GT6 Mk 3. However, for such a small car there was a great deal of stowage space behind the seats. Under the load floor were the petrol tank (left side) and the spare wheel (right side).

In a nutshell

In some ways the GT6 is a much misunderstood car. The Mk 1, agreed, had design flaws which were not cleared up until the Mk 2 was introduced; if the modern owner/restorer doesn't mind 'cheating' a little, by using non-authentic parts from later models, he can turn a Mk 1 into a much more satisfactory machine. Mk 2s and Mk 3s were excellent and well-developed little sports coupés.

Once you come to terms with the fact that the GT6 was purely a two-seater coupé – really no more than a faster, smoother, Spitfire with fastback styling – you see what an appealing package it offers today.

SPECIFICATIONS

	GT6 Mk 1	GT6 Mk 2
Engine	In-line six	In-line six
Construction	Cast-iron block, cast-iron head	Cast-iron block, cast-iron head
Main bearings	Four	Four
Bore/stroke	74.7 x 76mm	74.7 x 76mm
Capacity	1998cc	1998cc
Valves	Pushrod ohv	Pushrod ohv
Compression ratio	9.5:1	9.25:1
Fuel system	Two Zenith-Stromberg Type 150CD carburettors	Two Zenith-Stromberg Type 150CD carburettors
Power	95bhp (net) at 5000rpm	104bhp (net) at 5000rpm
Torque	117lb ft at 3000rpm	117lb ft at 3000rpm
Transmission	Four-speed manual, optional overdrive on top and third	Four-speed manual, optional overdrive on top and third
Top gear	20.1mph at 1000rpm in top when overdrive not fitted, 21.2mph at 1000rpm in overdrive top	20.1mph at 1000rpm in top when overdrive not fitted, 25.2 or 21.2mph at 1000rpm in overdrive top
Final drive	Hypoid, 3.27:1 ratio (without) 3.89:1 (with overdrive fitted)	Hypoid, 3.27:1 ratio (without overdrive); 3.27:1 or (later) 3.89:1 (with overdrive)
Brakes	9.7in Girling front discs and 8 x 1.25in rear drums	9.7in Girling front discs and 8 x 1.25in rear drums
Suspension, front	Ind. by coil springs, wishbones and anti-roll bar, telescopic dampers	Ind. by coil springs, wishbones and anti-roll bar, telescopic dampers
Suspension, rear	Ind. by transverse leaf spring, swing axles, telescopic dampers and radius arms	Ind. by transverse leaf spring, reverse lower wishbones, telescopic dampers and radius arms
Steering	Rack and opinion	Rack and pinion
Wheels/tyres	Steel disc wheels, 4.5in rims, or optional centre-lock wire spoke wheels, 4.5in rims. 155-13in tyres	Steel disc wheels, 4.5in rims, or optional centre-lock wire spoke wheels, 4.5in rims. 155-13in tyres
Chassis	Pressed steel backbone style chassis frame, with box section side members	Pressed steel backbone style chassis frame, with box section side members
Body	Pressed steel bodyshell, with two seats and two passenger doors. Fast back coupé/hatchback style. Entire nose section lifted up for engine bay/suspension access	Pressed steel bodyshell, with two seats and two passenger doors. Fastback coupé/hatchback style. Entire nose section lifted up for engine bay/suspension access

DIMENSIONS

	GT6 Mk 1	GT6 Mk 2
Length	12ft 1in	12ft 1in
Width	4ft 9in	4ft 9in
Height	3ft 11in	3ft 11in
Wheelbase	6ft 11in	6ft 11in
Unladen weight	1904lb	1904lb

GT6 Mk 3

In-line six

Cast-iron block, cast-iron head

Four

74.7 x 76mm

1998cc

Pushrod ohv

9.25 : 1

Two Zenith-Stromberg Type 150CD carburettors

104bhp (net) at 5000rpm

117lb ft at 3000rpm

Four-speed manual, optional overdrive on top and third

20.1mph at 1000rpm in top when overdrive not fitted, 21.2mph at 1000rpm in overdrive top

Hypoid, 3.27 : 1 ratio (without overdrive); 3.89 : 1 (with overdrive)

9.7in Girling front discs and 8 x 1.25in rear drums

Ind. by coil springs, wishbones and anti-roll bar, telescopic dampers

Ind. by transverse leaf spring, reverse lower wishbones, telescopic dampers and radius arms; from February 1973, by 'swing spring' transverse leaf spring, swing axles but no lower wishbones

Rack and opinion

Steel disc wheels, 4.5in rims, or optional centre-lock wire spoke wheels, 4.5in rims. 155-13in tyres

Pressed steel backbone style chassis frame, with box section side members

Pressed steel bodyshell, with two seats and two passenger doors. Fastback coupé/hatchback style. Entire nose section lifted up for engine bay/suspension access

12ft 5in

4ft 10.5in

3ft 11in

6ft 11in

2030lb

Production History

Spitfire

OCTOBER 1962	Original Spitfire model announced, with Michelotti-styled sports car body, backbone chassis, all-independent suspension, and 1147cc engine. Original UK retail price £730.
SEPTEMBER 1963	Overdrive (on top and third), plus detachable steel hardtop, introduced as options.
DECEMBER 1964	Mk 1 discontinued, and replaced by the slightly improved Mk 2, still with 1147cc engine. Original UK retail price £678. (UK purchase tax had been reduced since 1962.)
JANUARY 1967	Mk 2 discontinued, and replaced by Mk 3, complete with raised front bumper style, 1296cc. Original UK retail price £717.
1969 MODEL-YEAR	For USA cars *only,* a new-style facia.
OCTOBER 1969	Cosmetic style changes and new steering wheel for 1970, along with wider (4.5in) wheel rims.
NOVEMBER/ DECEMBER 1970	Mk IV production started in November, Mk 3 production discontinued in December. Mk IV with completely re-skinned body and cut-off tail, new facia (see USA for 1969), plus swing-spring rear suspension, all-synchromesh gearbox. New-style optional hardtop. Original UK price £962.
1973	For the USA, a 1493cc de-toxed engine. For all cars, two-inch rear track increase, reclining seats, fire-retardant upholstery, and related cosmetic changes. Centre-lock wire-spoke wheel option abandoned during year.
DECEMBER 1973	Front spoiler added under bumper.
DECEMBER 1974	Mk IV production discontinued, replaced by 1500 model, 1493cc, higher gearing. Original UK retail price £1509.
MARCH 1977	Style and cosmetic facelift.
AUGUST 1980	Last 1500 model built. Spitfire production ended at Canley after almost 18 years. Final UK retail price £5524.

GT6

OCTOBER 1966	GT6 Coupé launched, at same time as Vitesse 2-litre saloon, with which it shared engine and transmission. Effectively a Spitfire structure and style, with fastback/hatchback roof. Original UK retail price £985.
JULY/SEPTEMBER 1968	Mk 2 production began in July, Mk 1 production ended in September. Mk 2 with 104bhp (was 95bhp), and lower-wishbone rear suspension, plus style changes. Original UK price £1148. Mk 2 called 'GT6 Plus' for sale in USA.
OCTOBER 1969	Cosmetic style changes, reclining seats, and new steering wheel.
OCTOBER/DECEMBER 1970	Mk 3 production began in October, Mk 2 production ended in December. Mk 3 with re-skinned body style, cut-off tail, and many cosmetic style changes. Original UK retail price £1287. USA-version of Mk 3 now called Mk 3.
FEBRUARY 1973	Revisions for 1973 including same style changes as for 1973-model Spitfire, plus adoption of Spitfire 'swing-spring' rear suspension, and wide track. (Swing-spring production began in September 1972.
NOVEMBER/ DECEMBER 1973	Mk 3 withdrawn from production. Final UK retail price £1453.

Production Figures

By any standards, the Spitfire and GT6 were very successful cars. Throughout its long and distinguished career the Spitfire consistently outsold the rival Sprite/Midget series; because it was developed in such an easy, low-budget way, the GT6's less flamboyant career was also something of which Standard-Triumph (or, later, British Leyland) could be proud.

What cars were built at what time?

The dates quoted below are for actual start-up or close-down occasions, but *not* the month in which a car was officially put on sale, or the last cars were delivered. As an example, the British-market Spitfire Mk 3 was not revealed until March 1967, which was two months after assembly began, while the last of the USA-market 1500s was not sold until the late spring of 1981.

The first production-line Spitfires were built in October 1962, and the Mk 2 model took over in December 1964. The Mk 3 was phased in from January 1967, and the completely re-skinned Mk IV took over in November 1970, though some Mk 3s were also built in December of that year. The final derivative, the UK-market 1500, started life in December 1974, and the last car of all was produced in August 1980.

GT6 production started up in July 1966, and gradually gave way to the 'lower-wishbone' Mk 2 in the summer of 1968 (there was a two month overlap). The re-skinned Mk 3 entered production in October 1970, just a matter of days before the similarly re-skinned Spitfire Mk IV appeared. The last GT6s of all were built in November 1973, though quantity production had ended early in October 1973.

Spitfire Sales

Mk 1	45,573
Mk 2	37,409
Mk 3	65,320
Mk IV	70,021
1500*	95,829

Total Spitfire sales were 314,342 – an average achievement, over 18 years, of 17,500 cars a year. The best production year of all was 1964, when 23,387 cars were produced, and the least successful *full* year was 1979, when only 10,276 were produced.

A total of 139,547 cars were sold in the USA. The best-selling year of all was in 1978, when no fewer than 10,231 cars were sold – 48 per cent of Spitfire production for that period.

GT6 Sales

Mk 1	15,818
Mk 2	12,066
Mk 3	13,042

Total GT6 sales were 40,926 – an average achievement over rather more than seven years of about 5900 cars a year.

A total of 22,658 GT6s were sold in the USA – more than half of all cars built. In its best year, 1968, the GT6 recorded 4302 sales.

* This figure does not include 1493cc-engined versions of the Mk IV which were sold in the USA in 1973 and 1974.

At the painting stage all Triumph cars were routed down the same group of production lines. In this study we can see Spitfires, GT6s, Heralds, Vitesses, and Triumph 2000s.

At one stage of assembly at Triumph's large Canley plant, all the Herald family of cars were placed on the same line, which progressed the cars sideways towards completion. In this line-up, which dates from 1967 or 1968, there are Heralds, Spitfires and GT6s.

USA-Market Specifications

Starting in the late 1960s, United States legislators imposed ever more severe exhaust emission and safety laws on cars sold there. This meant that USA-spec. Spitfires and GT6s became more and more different from the home-market counterparts. As USA deliveries accounted for 44 per cent of all Spitfires, and 55 per cent of all GT6s, it is important that these differences should be listed.

Spitfire differences for the USA market

1962 to 1968	Minor details only
1969 (Mk 3)	8:5.1, 68bhp at 5500rpm; 73lb ft at 3000rpm; 1652lb
1970 (Mk 3)	As 1969, but 9:1, single Zenith-Stromberg carb
1971 (Mk IV)	As 1970, but 58bhp at 5200rpm; 72lb ft at 3000rpm
1972 (Mk IV)	As 1971, but 8:1, 48bhp at 5500rpm; 61lb ft at 2900rpm; axle ratio 4.11:1
1973 (Mk IV)	As 1972, but 1493cc, 7.5:1, 57bhp at 5000rpm; 74lb ft at 3000rpm; axle ratio 3.89:1
1974 (Mk IV)	As 1973, and 155-13in tyres
1975 (1500)	As 1974 Mk IV/1500: Weight 1828lb
1976 (1500)	9:1
1977 (1500)	7.5:1 and catalyst fitted
1978 (1500)	1850lb
1979 (1500)	Length 13ft 1.5in, due to use of energy-absorbing bumpers
1980	1875lb; no longer sold in California

GT6 differences for the USA market

1967 and 1968	Minor details only
1969 and 1970 (Mk 2 – called 'Plus' in USA)	Overdrive always with 3.89:1 final drive, 95bhp at 4700rpm; 117lb ft at 3400rpm
1971 (Mk 3 – same name used in USA)	9.25:1, 90bhp at 4700rpm; 116lb ft at 3400rpm
1972 and 1973 (Mk 3)	8.0:1, 79bhp at 4900rpm; 97lb ft at 2900rpm

RED BANK
IRPORT

Clubs

above
This was the scene in Holland where the Third International Spitfire weekend was held at Arcen in September 1985.

top left
The Spitfire was launched in the USA in 1963, and was sold in almost exactly the same guise as in Britain. The only discernible difference is the extra wheel trims.

bottom left
One of the last GT6s to be delivered to North America was this Mk 3, built with side indicator repeaters on the front and rear wings.

Spitfire and GT6 owners have the opportunity of joining several Triumph clubs, each of which has its own appeal. Effectively, these split into two groupings – Club Triumph and other clubs affiliated to it, and the Triumph Sports Six Club.

All are completely independent of what is now the Rover Group (ex-Leyland, ex-BL Cars, ex-Austin-Rover). However, some years ago, when British Leyland was trying to retain its dominance of the British motor industry Club Triumph was affiliated to the factory's public relations and publicity department, to the exclusion of other enthusiasts' organizations. However as British Leyland nose-dived, so did the corporation's interest in clubs and its motoring heritage. The result was that in the 1980s, the Triumph Sports Six Club, in particular, grew larger and much more active than the previously 'official' organization.

Triumph Sports Six Club

It all began with this small classified advert in *Thoroughbred & Classic Cars* and *Exchange & Mart*:

> **Triumph Owners** interested in forming a club for all vehicles using Herald chassis, Vitesse Courier van, Bond Equipe, etc, please contact Paul Swanson, 54 Caves Lane, Bedford.

Paul soon arranged a get-together at the THF Travelodge on M1 at Newport Pagnell. Thirteen people turned up on that occasion, chose the name of Triumph Sports Six Club for this enterprise, and it was the start of something big. Even though Paul later dropped out of the limelight, he had clearly sparked off a great deal of interest. Membership exceeded the 1000 mark by the beginning of the 1980s, passed 5000 in August 1984, and early in 1988 was well beyond the 11,000 mark; Spitfires and GT6s form a good proportion of those numbers. The growth continues, as well it might, for Club Manager Bill Sunderland estimates that there are 75,000 surviving Herald-family vehicles still in use in the UK, and probably as many in the USA.

The TSSC is a club which was founded, became established, and has prospered on a simple theme – it was dedicated to the preservation and restoration of every 'Herald-chassis vehicle' – which includes cars as different as a GT6 Mk 3 and a Herald-chassis special; along the way there has even been space for interest in the Herald-engined Amphicars as well! Not only, therefore, does the club accommodate many types of vehicles, but it has gone to some lengths to encourage the tracing of stocks of hard-to-find spares, and the re-manufacture of parts otherwise consigned to oblivion.

A simple newsletter became a fully fledged monthly magazine – now called *The Courier* – in 1979. This has evolved into a glossy and extremely attractively produced publication full of technical hints and tips, restoration articles and, as important as anything else, detailed adverts from restorers and parts suppliers. There is also the quarterly publication *Turning Circle*, which concentrates on a particular type of car, or a single subject.

President John Griffiths runs a highly modified Vitesse Convertible, Bill Sunderland runs a GT6 Mk 2, a Vitesse and a Spitfire for racing, while Trudi Squibbs runs a GT6 Mk I and has two more GT6s at home!

During 1988, members voted to change and up-date the constitution of the club, for its special motor insurance schemes had grown so rapidly that a new Financial Services division needed to be set up. By the time this book is published the TSSC will have six full-time staff at its headquarters, along with several insurance specialists working through the brokers, Footman James & Co. Ltd. That insurance scheme, by the way, is now used by more than 25 per cent of the membership, and not only provides Spitfire and GT6 owners with full comprehensive cover, but at an agreed value for their cars – something extremely difficult for any other one-make scheme to match.

There's no doubt that *every* Spitfire or GT6 owner ought to be a member of the TSSC. More details from:

Ms Trudi Squibbs,
Membership Secretary,
TSSC,
121B St Marys Road,
Market Harborough,
LEICS LE16 7DT (Tel: 0858-34424)

Club Triumph

Years ago the ancestors of this Club had links with the factory and it became something of an 'official' organization. At one time in the 1970s, indeed, the factory would only have dealings with other Triumph-orientated clubs if these had become associated to Club Triumph.

Today Club Triumph is a broad collection of centres, five of them, administered from the Coventry area by:

Malcolm Warren,
Club Triumph Ltd.,
14 John Simpson Close,
Wolston,
Coventry,
WARWICKS CV8 3HX (Tel: 0203-544770)

Each area acts as a separate organization, with regular social and competitive meetings, spares holdings, and technical advice centres. Each tends to build its own links with a local Rover Group dealer.

The largest area – the North London Group – has 600 members, there are another 500 in the Eastern (East Anglia-based) Group, and 400 in the Midlands, while the Club also has links with organizations like the Dolomite Sprint register (200 members), the Razor Edge Club, and the Stag Owners Club.

The Triumph Sporting Owners Club

In the beginning this was the original Spitfire Owners Club, and was based in the Manchester/North-West area. In recent years it changed its name, 'went national', opened up its general membership to Triumph-owning members, and now has seven local groups.

About half the current membership of 600 run Spitfires, and many of them are also members of the TSSC.

The Triumph Sports Six Club is much the largest enthusiasts' organization looking after the interests of Spitfire and GT6 owners. One of its most outstanding services is a glossily-produced monthly magazine called The Courier.

Further details from:

Richard King,
Chairman and Membership Secretary,
16 Windsor Road,
Hazel Grove,
Stockport,
CHESHIRE SK7 4SW (Tel: 061-456-0823)

Triumph Sports Car Club

This Bristol-based organization was originally a branch of Club Triumph, but has now adopted its own name. It is a small, geographically localized organization, with about 120 members at the time of writing. Like other clubs of this type, activities include monthly meetings, social events, and closed competitions.

Details from:

David Sloggett,
Membership Secretary,
104 St Peter's Rise,
Headley Park,
BRISTOL BS13 7NE

At this point I should also mention:

The British Motor Industry Heritage Trust

This is not an enthusiasts' club, in that individuals cannot join it, and it does not promote club nights, rallies or other annual fixtures. Based at Studley Castle, Studley, Warwicks B80 7AJ (Tel: 052785-4015/7337), it exists to maintain and develop a collection of vehicles, pictures, documents and other memorabilia of British motor cars; although its scope is widening all the time it has more ex-British Leyland archive material and cars than any other types.

Much of Standard-Triumph's production archive, covering the Spitfire/GT6 period, is now held by the BMIHT. Individual owners are able to get a 'Production Record Trace' of the original specification of their cars by contacting:

Anders Clausager,
Archivist,
The British Motor Industry Heritage Trust,
Castle Road,
Studley,
WARWICKS B80 7AJ (Tel: 052785-4015/7337)

Early in 1988 the fee for this service was raised to £5 (or $10) per car.

General Manager Richard Westcott told me that although the BMIHT has no financial or commercial links with any of the specialist Triumph clubs, it likes to try to support them in any way possible, including loaning cars, making its premises at Syon Park available for meetings, and helping to run events. It has also promoted the growth of the Association of Heritage Approved specialists, of which several Triumph restorers and parts suppliers are members.

Specialists

Because the Spitfire and GT6 sports cars have now been out of production for some considerable time, it is increasingly difficult to get spare parts and expertise through the retail motor trade. Standard-Triumph dealers have long ago been converted to Austin-Rover dealerships or – worse – have abandoned the sinking ship and started selling an entirely different make of car.

That is the discouraging news. The good news is that in recent years a number of specialist concerns, concentrating on the Herald-chassis family of cars, has developed; there is now a great deal of expertise, not only in the provision of existing parts and in matters of interchangeability, but in re-manufacture and in renovation.

The following is a selection of the specialists who are well known to the Triumph Sports Six Club.

Triumph Bookshop, Quad Marketing, Welland House, 9 Acorn Close, Lubenham, Market Harborough, Leics LE16 9SP (Tel: 0858-32110). This company not only produces the Triumph Sports Six Club's impressive publications, but also has a complete range of Spitfire/GT6 Handbooks, Parts Catalogues, and Workshop Manuals. In almost every case these have been reproduced from original factory items deleted years ago. In addition, there are stocks of books about the marque, and a selection of restoration guides.

John Kipping, Triumph Spares, 124 Aldermans Green Road, Coventry CV2 1PP (Tel: 0203-683926). This is one of the largest of the businesses specializing in Herald/Vitesse/Spitfire/GT6 parts. John Kipping was once the TSSC's Spares Officer, and has unrivalled expertise. He is at the forefront of the move to get obsolete parts remanufactured. The company specializes in gearbox, overdrive and differential rebuilds.

John Hill's Ltd, Spitfire-Herald Centre, Arthur Street, Redditch, Worcs B98 8JY (Tel: 0527-20880). Not only is this company the largest of all the stockists dealing in Spitfires and GT6s, but it is the only one to be a British Motor Heritage Approved supplier.

Triumphtune (UK) Ltd, 22-28 Manor Road, Richmond, Surrey (Tel: 01-948-6668). Terry Hurrell, son of the legendary Syd Hurrell of Triumph TR fame, runs this operation, which is linked to the Cox & Buckles company (whose speciality is Triumph TRs). Not only does this company provide a wide range of replacement parts, but also offers tuning components for engines, transmissions and suspension systems.

Newton Commercial, Eastlands Industrial Estate, Leiston, Suffolk IP16 4LL (Tel: 0728-832880). This company specializes in the re-manufacture of trim for Spitfires. All items are made to original design and colour specifications. Available components include door and seat trims, tunnel trim, and panels around the dash, parcel tray, and boot.

Cambridge Triumph Spares, 60 Ramsey Road, St Ives, Huntington, Cambs PE17 6XL (Tel: 0480-301671). Mainly bodyshell and related items, including rubber, decoration, badging.

Classic Automotive, St Michaels Way, Middlewich, Cheshire CW10 9DX (Tel: 0606-846474/5). Concentrates on supply of parts for brakes, steering and suspension, with many other general spares held in stock.

Accessory and Motor Supplies, Unit A8, Faraday Road, London Road Industrial Estate, Newbury, Berks (Tel: 0635-44669). A full range of mechanical components and body panels.

Chic Doig, Unit 1, West Mills, Mill Street, Kirkcaldy, Fife, Scotland (Tel: 0592-206439). Produces many spares by breaking otherwise hopeless Herald-family cars, is now involved in providing refurbished chassis frames, and offers body panels and stainless steel exhaust systems.

One of the leading Spitfire/GT6 specialists is John Kipping, whose catalogue details the hundreds of mechanical, chassis and body parts which can be supplied to restore a car to health. Here, and on the next page, is just a selection – showing that an owner can buy almost anything to rebuild his car.

Chassis
SPITFIRE/GT6

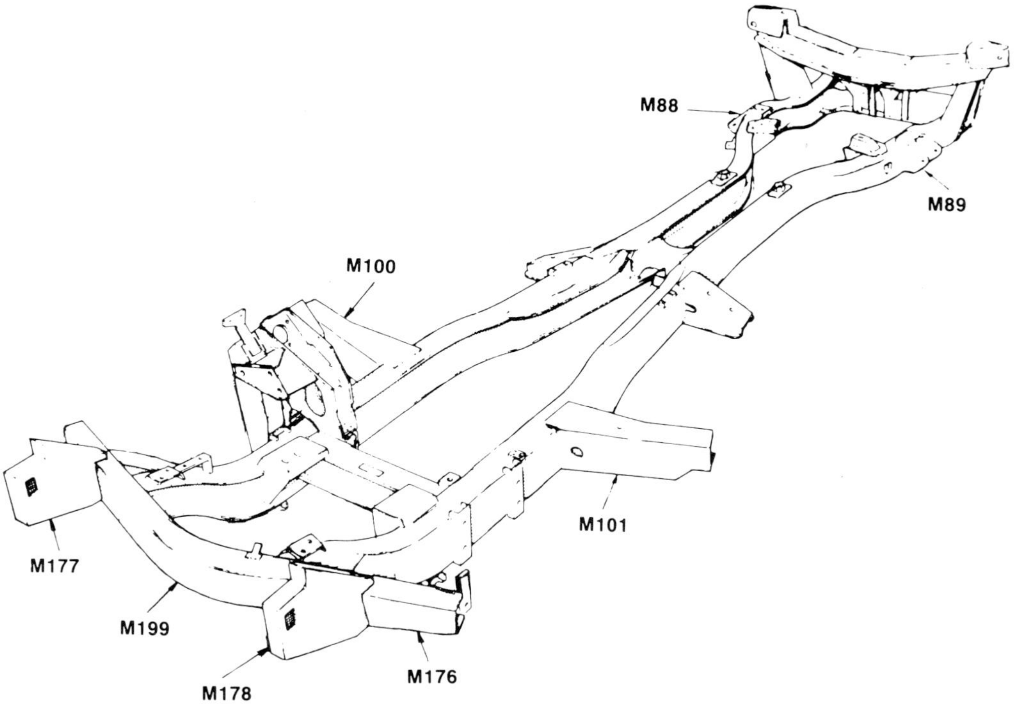

Bonnet Assembly
SPITFIRE I/II/III/GT6 I/II

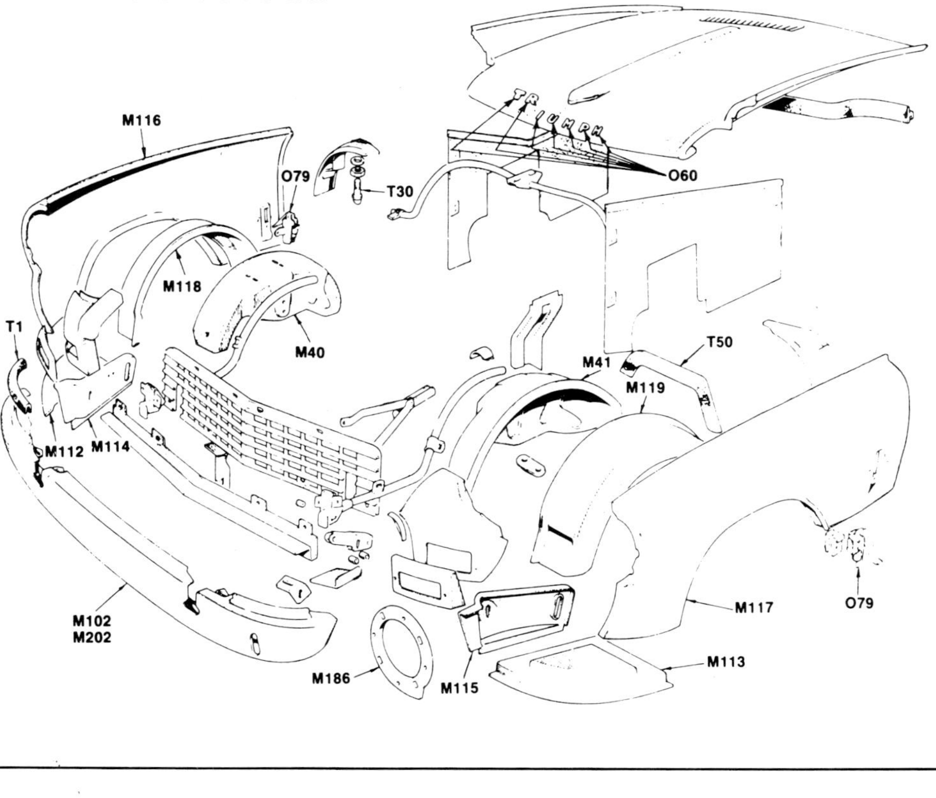

British Sports Car Centre, 299-309 Goldhawk Road, London W12 8EZ (Tel: 01-748-7823, or 01-741-7075). Offers a whole range of mechanical, and some body, parts.

Spitfires (UK), Unit 11A, Branston House, West Avenue, Wigston, Leicester (Tel: 0533-812129). Probably the longest-established specialist in Spitfire/GT6 parts. Has had a rather 'bumpy' commercial career, but seems to be working well again. Particularly offers body repair panels and interior trim, plus a selection of mechanical parts.

GB Sports Cars, Freemans Depot, Stanbridge Road, Billington, Leighton Buzzard, Beds (Tel: 0525-378078), Deals mainly in secondhand Spitfire and GT6 parts, found by concentrating on the dismantling of worn-out cars.

Andy Rose, 103 Horsa Road, Northumberland Heath, Kent (Tel: 0322-338227). Deals mainly in servicing, tuning and repairs.

Vintage and Classic International Ltd, Unit 43B, Hartlebury Trading Estate, Kidderminster, Worcs DY10 4JB (Tel: 0299-251353). This company has set out to stock obsolete Lucas electrical items, and to re-manufacture Lucas electrical components using original tooling and equipment.

Honeybourne Mouldings, Eclipse Trading Estate, Birmingham Road, Alcester, Warwicks B49 5EQ (Tel: 0789-762071). As the title suggests, this company supplies hardtops in glass-fibre.

Midland Spitfires, 105 Brearley Street, Newtown, Birmingham, West Midlands B19 3XJ (Tel: 021-359-3315). This company is in modern premises close to the infamous M6 'Spaghetti Junction', and concentrates on bodyshell and fittings of all types.

L & M Motors, Rear of 82/86 Southend, Croydon, Surrey (Tel: 01-688-1622). Concentrating on restoration and repair/replacement of panelwork, rather than holding big stocks of mechanical parts.

John Wooley, Kiles Nest Workshop, Bulley Lane, Churcham, near Gloucester, Glos (Tel: 0452-79585). Provides a complete vehicle restoration and renovation service.

S.W. Classics, Brook House, Dartington, Totnes, Devon (Tel: 0803-865842). Offers a full body and mechanical repair service. Complete restorations are a speciality.

Floor Assembly SPITFIRE/GT6

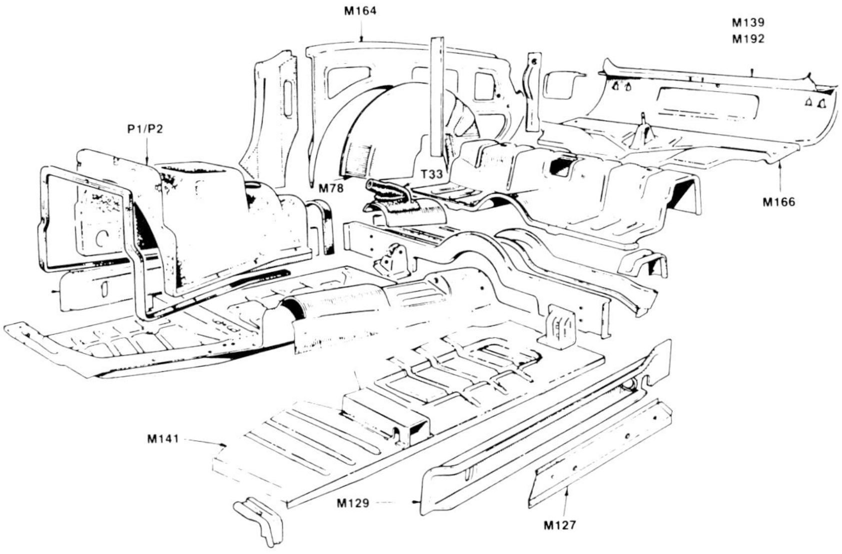

Body Panels SPITFIRE/GT6

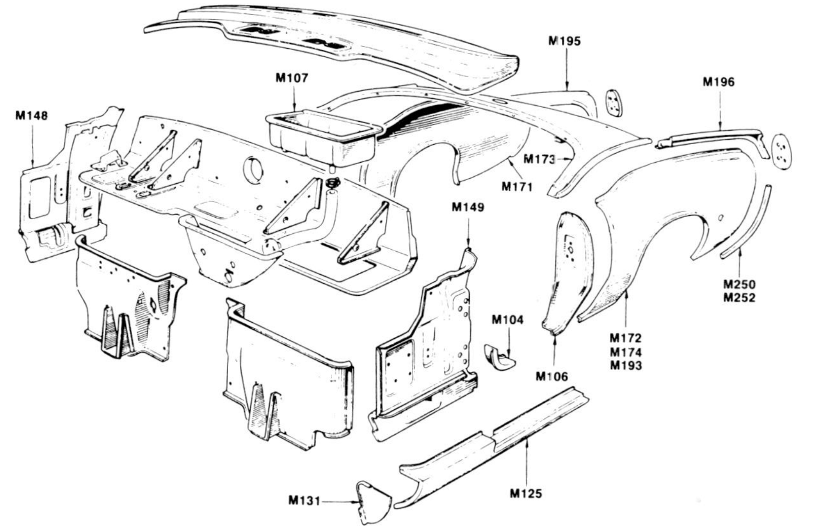

Chiswick Car Craft, Turnham Green Terrace Mews, Chiswick, London W4 1QU (Tel: 01-994-3395). This company, previously called Budget Welding, specializes in restoration of older cars, and concentrates on chassis and suspension rebuilds.

Old Parts for New

Mainly through the efforts of the Triumph Sports Six Club, and because the Spitfire and GT6s are now well-established classic cars, in general the parts supply situation is good. There may be occasions, however, when someone restoring one of the cars may have to look elsewhere for his parts.

In the 1960s and 1970s Standard-Triumph liked to get the best possible value out of its 'building blocks' – engines, transmissions, suspension parts, and other such details. For that reason, much of the Spitfire/GT6 running gear has also been used, in one form or another, in other Standard-Triumph models.

In every case, you should consult appropriate Parts Lists, to see if part numbers coincide. Even so, although there may be important differences, the following information will be useful.

Bodyshells and fittings

As with the chassis frames, the bodyshells themselves have nothing in common with those used in any other Standards or Triumphs.

Some electrical fittings such as headlamps, stop-tail lamps, and reverse lamps were standard 'off-the-shelf' Lucas fittings. These, though no longer available from Lucas, are still available in re-manufactured form, and (I understand) using original Lucas tools and moulds. Adverts for such items regularly appear in the 'classic' press.

The tail lamps used on 1970s Spitfires and GT6s were also used on the Stag of the period. However, those used on cars like the TR6 and the Triumph 2000/2.5 PI saloons were completely different.

On pre-reskin models, the Triumph front badge and the T R I U M P H letters were all standard items, to be found, from time to time, on 1960s Michelotti-style TRs and the Herald/Vitesse family.

The Spitfire/GT6 windscreen was the same as that of the TR4-TR6 sports car family, while the Spitfire's snap-action fuel filler cap was the same as that used in contemporary TRs.

Chassis frames

In this case, the Spitfire/GT6 frames were unique, and there is no point in looking at, say, Herald/Vitesse frames for help in restoring a car. However the front suspension 'towers', the pressed/welded assemblies supporting the spring/damper units, and the top wishbones, were common.

Engines

Spitfire

The three *basic* engines used in Spitfires – 1147cc, 1296cc and 1493cc – were those widely used in cars as different as the Herald 1200 and the Triumph 1500TC. However, although major parts like cylinder blocks, front end timing gear, cranks and connecting rods were usually common, there were several different cylinder heads, camshafts, piston types and, of course, carburation and manifolding.

However, it is worth knowing that the Herald 1200 and 12/50 models (1147cc) were in production from 1961 to 1970, and that the 13/60 models (1296cc) were in production from 1967 to 1971. The 1296cc engine was used in Triumph 1300s, Toledos and Dolomite 1300s from 1965 to 1980; in one form at least – the 1300TC of 1967-1970 – the engine was very similar indeed to that used in the Spitfire of the period. The 1493cc engine was used in the Triumph 1500/Toledo 1500/Triumph 1500TC/Dolomite 1500 family of saloons.

Although the 1300/1300TC/1500 models had front-wheel-drive, the engines themselves were very little different from more conventional front-engine/rear-drive types. The 1300TC was the first saloon to use the eight-port cylinder head also fitted to the Mk 3 Spitfire, though this was not used on the 1300 model; the eight-port head was commonized when the next generation Toledo/1500 models arrived in 1970.

Most important of all, however, is the fact that the MG Midgets of 1974-1979 (carrying chassis prefixes GAN6 . . .) used virtually the same engines as the Triumph Spitfire 1500s; the major difference, apparently, was in the exhaust manifold area. This was deliberate product planning by British Leyland, to commonize the cars for sale in the USA. Since Midgets are notorious for rusting away much faster than the Spitfire (there was no separate chassis frame on the Midgets), this means that some cars crumble away well before their engines are worn out!

GT6

The six-cylinder engine used in GT6s started life as a 1998cc unit in 1960 by being used in the Standard Vanguard Luxury Six, and the same engine was then adopted for use in the Triumph 2000 models. The Vanguard Six was penalized by the fitment of two semi-downdraught Solex 32 PHI carburettors which never found a home on any other make of car, though the 2000 used twin side-draught Zenith-Stromberg units. From late 1966 the engine was also pressed into use for the Vitesse 2-litre saloons and convertibles.

(The six-cylinder 1596cc engine fitted to the first-generation Vitesses of 1962-1966 is not really relevant here, for it used a small-bore cylinder block, with an entirely different cylinder block casting. From 1962 to 1965, too, this engine used the same type of semi-downdraught Solex carbs as the Vanguard Six.)

In 1967 the same engine was considerably uprated in one form, when it was enlarged to 2498cc, with the same bore but a much longer stroke, and at the same time an entirely new and more efficient 'full-width' type of cylinder-head casting was adopted. For the next decade this unit was used in cars as diverse as the TR5/TR250/TR6 sports cars and the 2.5PI/2500TC/2500S saloon cars.

The Mk I GT6's engine was rated at 95bhp, just like that of the Vitesse 2-litre of 1966-1968, and was virtually identical to that engine; it also had a lot in common with the Triumph 2000's engine, and used the same basic block and cylinder head as those cars.

From late 1968, the Mk 2 and Mk 3 GT6s used the rationalized six-cylinder block (the same for 2.0-litre and 2.5-litre engines), and the new 'full-width' cylinder heads, all of which were used in late-model 2000s and 2.5s.

(As any GT6 enthusiast knows, the installation of 2498cc engines, even fitted with Lucas fuel injection, is really quite a straightforward business, as the engine is physically no longer, or wider, than the 1998cc engine which it displaces.)

Gearboxes

Spitfire

Except for very minor detail, the gearbox used in the Spitfire 1, 2 and 3 models was the same as that used in the Herald 1200 and 13/60 models of the period – which is to say that it was wide-ratio type without synchromesh on first gear. The casing, but not the internal ratios or the gear wheels themselves, was also shared with the Vitesse 1600 of the 1962-1966 period. The optional overdrive was effectively that fitted to the Vitesses of the period – overdrive was never offered on the Herald range, though there would have been space to fit it, if necessary.

The gearbox fitted to the Vitesse 1600 used the same casing, selector mechanisms, and general layout as that of the 1962-1970 Spitfire (along with the Herald 1200 and 13/60 models), but it had an entirely different, closer, set of internal ratios. These ratios in fact were homologated for the Spitfire in 1964, and used in rallying in 1964.

Throughout the 1970s, for the Mk IV and 1500 models, the Spitfire used the same all-synchromesh rationalized gearbox design (but not necessarily the same internal ratios) as also found in cars like the Mk 3 GT6, the Triumph 1500TC/Dolomite 1300/Dolomite 1500/Dolomite 1850 range, the Morris Marina, and of course the Triumph-engined MG Midget of 1974-1979. Single-rail selection was adopted for the Spitfire 1500 (and MG Midget 1500); it had already featured on the Marina from 1971.

GT6

Specifically for this model and for the Vitesse 2-litre of 1966-1971, Triumph developed an all-synchromesh four-speed gearbox. Although its design 'roots' were the same as those of the Vitesse 1600, the gear train was longer, and there were many other detailed differences.

To all intents and purposes, the gearbox and (optional) overdrive of the Vitesse 2-litre was identical to that used in the GT6 models. It was this gearbox which formed the basis of the Spitfire IV/1500 transmission, and of all the other cars in the British Leyland rationalized transmission range, already mentioned above.

Note: The gearbox/overdrive units fitted to the Triumph TRs, 2000/2.5s, Stags, and Dolomite Sprints were of an entirely different, larger and more robust type. Although such boxes *were* fitted to the 1964 Spitfire Le Mans

A study in Spitfire noses – the TSSC National Triumph rally, held in August 1985.

cars, they cannot, and should not, be mated to any ordinary road-going Spitfire.

Rear axles

As has already been made clear in the descriptive chapters, the Spitfires and GT6s used several different derivatives of the same type of chassis-mounted rear axle/differential. These assemblies were also fitted to the 'sister' cars – the Heralds, the Vitesse 1600 and Vitesse 2-litre models, but to no other Triumphs, nor to any other contemporary British Leyland model. The chassis-mounted rear axle fitted to TRs from the TR4A, to 2000s and 2.5s, and to the Stag, was a different, physically larger and more robust assembly.

Front suspension and steering

The amazing steering lock, allied to the wide-based wishbones which made it all possible, was one of the advanced technical features new to the Herald in 1959. In more developed form this front suspension assembly was fitted to every Herald, Vitesse, Spitfire and GT6 ever made. In much modified form, it was also fitted to the 1960s-design Lotus Elan, Elan Plus 2, and Europa models; these cars were progressively changed during their lives, and did not finally disappear until 1975.

The entire suspension was made in Hemel Hempstead by Alford & Alder, which was a subsidiary of Standard-Triumph. The vertical-link casting which supported the upper ball joint and the lower trunnion, and which held the front hub, seemed to be exactly the right size for several single-seater racing cars of the day, even for a few Formula 1 cars. This information, of course, is quite useless to those searching for spare parts for a Spitfire/GT6 rebuild!

When the Herald/Vitesse family died out, it was effectively replaced by the Toledo/1500TC/Dolomite series, which had entirely different front suspension components.

Author's Acknowledgments

I first met the Triumph Spitfire as a prototype in 1961, and tested several production cars for *Autocar* in the 1960s, so a complete list of those who have given me help along the way would be extremely long. I might even be accused of name dropping!

In connection with this book, however, I am particularly grateful to Anders Clausager of the British Motor Industry Heritage Trust, to Bill Sunderland and Trudi Squibbs of the Triumph Sports Six Club, to Derek Pollock of Club Triumph (North London), to John Kipping, Mike Kennoy and John Wooley (who are all preservation experts), and to Ralph Hardwick for helping me with photographic matters.

Publishers' Acknowledgments

For providing cars for photography (in various states of decay or restoration) the publishers wish to thank Tony Butterfield and Dave Boswell of Sports and Classics in Torrington, North Devon, and Matthew Symons of Bideford.